MAGNETISM
with paper clips,
magnets,
pins
and simple things

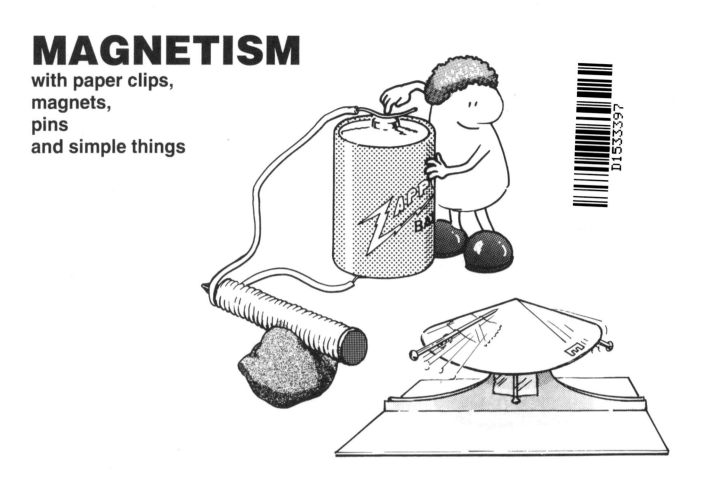

SCIENCE WITH SIMPLE THINGS SERIES

Conceived and
written by
RON MARSON

Illustrated by
PEG MARSON

TOPS LEARNING SYSTEMS

10970 S Mulino Road
Canby OR 97013
Website: topscience.org
Fax: 1(503) 266-5200

Oh, those pesky COPYRIGHT RESTRICTIONS !

Dear Educator,

TOPS is a nonprofit organization dedicated to educational ideals, not our bottom line. We have invested much time, energy, money, and love to bring you this excellent teaching resource.

And we have carefully designed this book to run on simple materials you already have or can easily purchase. If you consider the depth and quality of this curriculum amortized over years of teaching, it is dirt cheap, orders of magnitude less than prepackaged kits and textbooks.

Please honor our copyright restrictions. We are a very small company, and book sales are our life blood. When you buy this book and use it for your own teaching, you sustain our publishing effort. If you give or "loan" this book or copies of our Activity Sheets to other teachers, with no compensation to TOPS, you squeeze us financially, and may drive us out of business. Our well-being rests in your hands.

What if you are excited about the terrific ideas in this book, and want to share them with your colleagues? What if the teacher down the hall, or your homeschooling neighbor, is begging you for good science, quick! We have suggestions. Please see our *Purchase and Royalty Options* below.

We are grateful for the work you are doing to help shape tomorrow. We are honored that you are making TOPS a part of your teaching effort. Thank you for your good will and kind support.

Sincerely, Ron Marson

Purchase and Royalty Options:

Individual teachers, homeschoolers, libraries:

PURCHASE option: If your colleagues are asking to borrow your book, please ask them to read this copyright page, and to contact TOPS for our current catalog so they can purchase their own book. We also have an **online catalog** that you can access at www.topscience.org.

If you are reselling a **used book** to another classroom teacher or homeschooler, please be aware that this still affects us by eliminating a potential book sale. We do not push "newer and better" editions to encourage consumerism. So we ask seller or purchaser (or both!) to acknowledge the ongoing value of this book by sending a contribution to support our continued work. Let your conscience be your guide.

Honor System ROYALTIES: If you wish to make copies from a library, or pass on copies of just a few activities in this book, please calculate their value at 50 cents (25 cents for homeschoolers) per lesson per recipient. Send that amount, or ask the recipient to send that amount, to TOPS. We also gladly accept donations. We know life is busy, but please do follow through on your good intentions promptly. It will only take a few minutes, and you'll know you did the right thing!

Schools and Districts:

You may wish to use this curriculum in several classrooms, in one or more schools. Please observe the following:

PURCHASE option: Order this book in quantities equal to the number of target classrooms. If you order 5 books, for example, then you have unrestricted use of this curriculum in any 5 classrooms per year for the life of your institution. You may order at these quantity discounts:

02-09 copies: 90% of current catalog price + shipping.

10+ copies: 80% of current catalog price + shipping.

ROYALTY option: Purchase 1 book *plus* photocopy or printing rights in quantities equal to the number of designated classrooms. If you pay for 5 Class Licenses, for example, then you have purchased reproduction rights for any 5 classrooms per year for the life of your institution.

01-09 Class Licenses: 70% of current book price per classroom.

10+ Class Licenses: 60% of current book price per classroom.

Workshops and Training Programs:

We are grateful to all of you who spread the word about TOPS. Please limit duplication to only those lessons you will be using, and collect all copies afterward. No take-home copies, please. Copies of copies are prohibited. Ask us for a free shipment of as many current **TOPS Ideas** catalogs as you need to support your efforts. Every catalog contains numerous free sample teaching ideas.

ISBN 0-941008-54-1

CONTENTS

PART I — PREPARATION AND SUPPORT

PART II — ACTIVITIES AND LESSON NOTES

PART III — SUPPLEMENTARY PAGES

A TOPS Teaching Model

If science were only a set of explanations and a collection of facts, you could teach it with blackboard and chalk. You could require students to read chapters in a textbook, assign questions at the end of each chapter, and set periodic written exams to determine what they remember. Science is traditionally taught in this manner. Everybody studies the same information at the same time. Class togetherness is preserved.

But science is more than this. It is also process — a dynamic interaction of rational inquiry and creative play. Scientists probe, poke, handle, observe, question, think up theories, test ideas, jump to conclusions, make mistakes, revise, synthesize, communicate, disagree and discover. Students can understand science as process only if they are free to think and act like scientists, in a classroom that recognizes and honors individual differences.

Science is both a traditional body of knowledge and an individualized process of creative inquiry. Science as process cannot ignore tradition. We stand on the shoulders of those who have gone before. If each generation reinvents the wheel, there is no time to discover the stars. Nor can traditional science continue to evolve and redefine itself without process. Science without this cutting edge of discovery is a static, dead thing.

Here is a teaching model that combines both the content and process of science into an integrated whole. This model, like any scientific theory, must give way over time to new and better ideas. We challenge you to incorporate this TOPS model into your own teaching practice. Change it and make it better so it works for you.

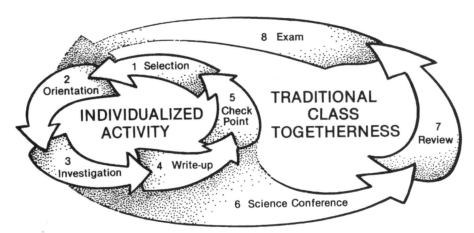

1. SELECTION

Students generally select activity pages in sequence, because new concepts build on old ones in a specific order. There are, however, exceptions to this pattern: students might skip a lesson that is not challenging; repeat an activity with doubtful results; add an experiment to answer their own "what-would-happen-if?" questions.

Working at their own pace, students fall into a natural routine that creates stability and order. They still have questions and problems, to be sure, but remain purposefully engaged with a definite sense of direction.

2. ORIENTATION

Any student with basic reading skills can successfully interpret our carefully designed activity page directions. If your class is new to TOPS, it may take a while for your students to get used to following directions by themselves, and to trust in their own problem-solving ability.

When students ask you for help, first ask them to read what they don't understand. If they didn't read the instruction in the first place, this should clear things up. Identify poor readers in your class. Whey they ask, "What does this mean?" they may be asking in reality, "Will you please read these directions aloud?"

Beyond reading comprehension, certain basic concepts and skills may also be necessary to complete some activity sheets. You can't, for example, expect students to measure the length of something unless they know how to use a ruler as well. Anticipate and teach prerequisite concepts and skills (if any) at the beginning of each class period, before students begin their daily individualized work. Different age groups will require different levels of assistance: primary students will need more introductory support than middle school students; secondary students may require none at all.

3. INVESTIGATION

Students work through the activity pages independently and cooperatively, They follow their own experimental strategies and help each other. Encourage this behavior by helping students only after they have tried to help themselves. As a resource teacher, you work to stay out of the center of attention, responding to student questions rather than posing teacher questions.

Some students will progress more rapidly than others. To finish as a cohesive group, announce well in advance when individualized study will end. Expect to generate a frenzy of activity as students rush to meet your deadline. While slower students finish those core activities you specify, challenge your more advanced students with Extension activities, or to design original experiments.

4. WRITE-UP

Activity pages ask students to explain the how and why of things. Answers may be brief and to the point, with the exception of those that require creative writing. Students may accelerate their pace by completing these reports out of class.

Students may work alone, or in cooperative lab groups. But each one should prepare an original write-up, and bring it to you for approval. Avoid an avalanche of write-ups near the end of the unit by enforcing this simple rule: each write-up must be approved before starting the next activity.

5. CHECK POINT

Student and teacher together evaluate each write-up on a pass/no-pass basis. Thus no time is wasted haggling over grades. If the student has made reasonable effort consistent with individual ability, check off the completed activity on a progress chart. Students keep these in notebooks or assignment folders kept on file in class.

Because the student is present when you evaluate, feedback is immediate and effective. A few moments of your personal attention is surely more effective than tedious margin notes that students may not heed or understand. Remember, you don't have to point out every error. Zero in on particular weaknesses. If reasonable effort is not evident, direct students to make specific improvements and return for a final check.

A responsible lab assistant can double the amount of individual attention each student receives. If he or she is mature and respected by your students, have the assistant check even-numbered write-ups, while you check the odd ones. This will balance the work load and assure equal treatment.

6. SCIENCE CONFERENCE

Individualized study has ended. This is a time for students to come together, to discuss experimental results, to debate and draw conclusions. Slower students learn about the enrichment activities of faster classmates. Those who did original investigations or made unusual discoveries share this information with their peers, just like scientists at a real conference.

This conference is an opportunity to expand ideas, explore relevancy and integrate subject areas. Consider bringing in films, newspaper articles and community speakers. It's a meaningful time to investigate the technological and social implications of the topic you are studying. Make it an event to remember.

7. REVIEW

Does your school have an adopted science textbook? Do parts of your science syllabus still need to be covered? Now is the time to integrate traditional science resources into your overall program. Your students already share a common background of hands-on lab work. With this base of experience, they can now read the text with greater understanding, think and problem-solve more successfully, communicate more effectively.

You might spend just a day here, or an entire week. Finish with a review of major concepts in preparation for the final exam. Our review/test questions provide an excellent resource for discussion and study.

8. EXAM

Use any combination of our review/test questions, plus questions of your own, to determine how well students have mastered the concepts they've been learning.

Now that your class has completed a major TOPS learning cycle, it's time to start fresh with a brand new topic. Those who messed up and got behind don't need to stay there. Everyone begins the new topic on an equal footing. This frequent change of pace encourages your students to work hard, to enjoy what they learn, and thereby grow in scientific literacy.

Getting Ready

Here is a checklist of things to think about and preparations to make before beginning your first lesson on MAGNETISM.

✔ Review the scope and sequence.

Take just a few minutes, right now, to page through all 20 lessons. Pause to read each *Objective* (top left column of the Teaching Notes) and scan each lesson.

✔ Set aside appropriate class time.

Allow an average of perhaps 1 class period per lesson (more for younger students), plus time at the end of this module for discussion, review and testing. If you teach science every day, this module will likely engage your class for about 4 weeks. If your schedule doesn't allow this much science, consult the logic tree on page E to see which activities you can safely omit without breaking conceptual links between lessons.

✔ Number your activity sheet masters.

The small number printed in the top right corner of each activity page shows its position within the series. If this ordering fits your schedule, copy each number into the blank parentheses next to it. Use pencil, as you may decide to revise, rearrange, add or omit lessons the next time you teach this topic. Insert your own better ideas wherever they fit best, and renumber the sequence. This allows your curriculum to adapt and grow as you do.

✔ Photocopy sets of student activity sheets.

Supply 1 per student, plus supplementary pages as required. Store these in manila folders for convenient access by your students. Please honor our copyright notice at the front of this book. We allow you, the purchaser, to photocopy all permissible materials, as long as you limit the distribution of copies you make to the students you personally teach. Encourage other teachers who want to use this module to purchase their own books. This supports TOPS financially, enabling us to continue publishing new titles for you.

✔ Collect needed materials.

See page D, opposite, for details.

✔ Organize a way to track assignments.

Keep student work on file in class. If you lack a file cabinet, a box with a brick will serve. File folders or notebooks both make suitable assignment organizers. Students will feel a sense of accomplishment as they see their folders grow heavy, or their notebooks fill, with completed assignments. Since all papers stay together, reference and review are facilitated.

Ask students to number a sheet of paper from 1 to 20 and tape it inside the front cover of their folders or notebooks. Track individual progress by initialing lesson numbers as daily assignments pass your check point.

✔ Review safety procedures.

In our litigation-conscious society, we find that publishers are often more committed to protecting themselves from liability suits than protecting students from physical hazards. Lab instructions are too often filled with spurious advisories, cautions and warnings that desensitize students to safety in general. If we cry "Wolf!" too often, real warnings of present danger may go unheeded.

At TOPS we endeavor to use good sense in deciding what students already know (don't stab yourself in the eye) and what they should be told (don't look directly at the sun.) Pointed scissors, pins and such are certainly dangerous in the hands of unsupervised children. Nor can this curriculum anticipate irresponsible behavior or negligence. As the teacher, it is ultimately your responsibility to see that common-sense safety rules are followed; it is your students' responsibility to respect and protect themselves and each other.

✔ Communicate your grading expectations.

Whatever your grading philosophy, your students need to understand how they will be assessed. Here is a scheme that counts individual effort, attitude and overall achievement. We think these three components deserve equal weight:

• Pace (effort): Tally the number of check points and extra credit experiments you have initialed for each student. Low ability students should be able to keep pace with gifted students, since write-ups are evaluated relative to individual performance standards on a pass/ no-pass basis. Students with absences, or those who tend to work slowly, might assign themselves more homework out of class.

• Participation (attitude): This is a subjective grade, assigned to measure personal initiative and responsibility. Active participators who work to capacity receive high marks. Inactive onlookers who waste time in class and copy the results of others receive low marks.

• Exam (achievement): Activities point toward generalizations that provide a basis for hypothesizing and predicting. The Review/Test questions beginning on page G will help you assess whether students understand relevant theory and can apply it in a predictive way.

Gathering Materials

Listed below is everything you'll need to teach this module. Buy what you don't already have from your local supermarket, drugstore or hardware store. Ask students to bring recycled materials from home.

Keep this classification key in mind as you review what's needed.

general on-the-shelf materials:	special in-a-box materials:
Normal type suggests that these materials are used often. Keep these basics on shelves or in drawers that are readily accessible to your students. The next TOPS module you teach will likely utilize many of these same materials.	*Italic type suggests that these materials are unusual. Keep these specialty items in a separate box. After you finish teaching this module, label the box for storage and put it away, ready to use again.*
(substituted materials):	*optional materials:
Parentheses enclosing any item suggests a ready substitute. These alternatives may work just as well as the original. Don't be afraid to improvise, to make do with what you have.	An asterisk sets these items apart. They are nice to have, but you can easily live without them. They are probably not worth an extra trip to the store, unless you are gathering other materials as well.

Everything is listed in order of first use. Start gathering at the top of this list and work down. Ask students to bring recycled items from home. The *Teaching Notes* may occasionally suggest additional *Extensions*. Materials for these optional experiments are listed neither here nor under *Materials*. Read the extension itself to determine what new items, if any, are required.

Quantities depend on how many students you have, how you organize them into activity groups, and how you teach. Decide which of these 3 estimates best applies to you, then adjust quantities up or down as necessary:

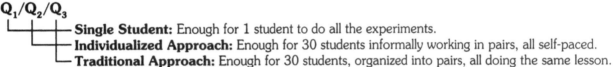

$Q_1/Q_2/Q_3$

— **Single Student:** Enough for 1 student to do all the experiments.
— **Individualized Approach:** Enough for 30 students informally working in pairs, all self-paced.
— **Traditional Approach:** Enough for 30 students, organized into pairs, all doing the same lesson.

KEY:	*special in-a-box materials* (substituted materials)	general on-the-shelf materials *optional materials

$Q_1/Q_2/Q_3$

2/60/60	ceramic magnets – see notes 1	5/75/75	*feet copper or aluminum wire, about 24 gauge, no thicker than 22 gauge, with soft easily stripped insulation.*
1 box	paper clips		
1 box	steel straight pins, 1 inch	1/15/15	size D dry cells
1 roll	aluminum foil	9/36/135	iron washers
8/40/120	pennies	2/20/30	wooden clothespins
1 spool	thread	1 package	clay
1 roll	masking tape	1/5/15	jars or equivalent – see notes 18
1 roll	clear tape	2/10/30	rubber bands
1/10/15	pairs of scissors	8/40/120	*cm thin bare copper or aluminum wire, about 30 or 32 gauge – see note 19*
1/1/3	paper punch tools		
1/10/15	styrofoam cups, small size	1/3/8	manila folders (pressed cardboard)
2/20/30	3x5 inch index cards	2/12/30	clothes hangers (stiff flat rulers or equivalent)
2/30/30	*plastic cups or equivalent – see notes 5*	1 package	uncooked rice grains
1/1/1	black permanent marker	1 box	staples
1/15/15	6½ cm (2½ inch) iron nails		

Sequencing Activities

This logic tree shows how all the activities in this book tie together. In general, students begin at the trunk of the tree and work up through the related branches. Lower level activities support the ones above.

You may, at your discretion, omit certain activities or change their sequence to meet specific class needs. However, when leaves open vertically into each other, those below logically precede those above, and should not be omitted.

When possible, students should complete the activities in the same sequence as numbered. If time is short, however, or certain students need to catch up, you can use this logic tree to identify concept-related horizontal activities. Some of these might be omitted, since they serve to reinforce learned concepts rather than introduce new ones.

For whatever reason, when you wish to make sequence changes, you'll find this logic tree a valuable reference. Parentheses in the upper right corner of each activity page allow you total flexibility. They are blank so you can pencil in sequence numbers of your own choosing.

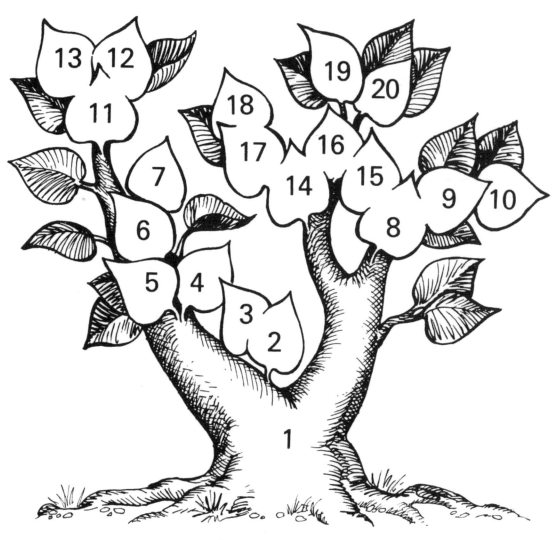

MAGNETISM 33
E

Gaining a Whole Perspective

Science is an interconnected fabric of ideas woven into broad and harmonious patterns. Use extension ideas in the teaching notes plus the outline presented below to help your students grasp the big ideas — to appreciate the fabric of science as a unified whole.

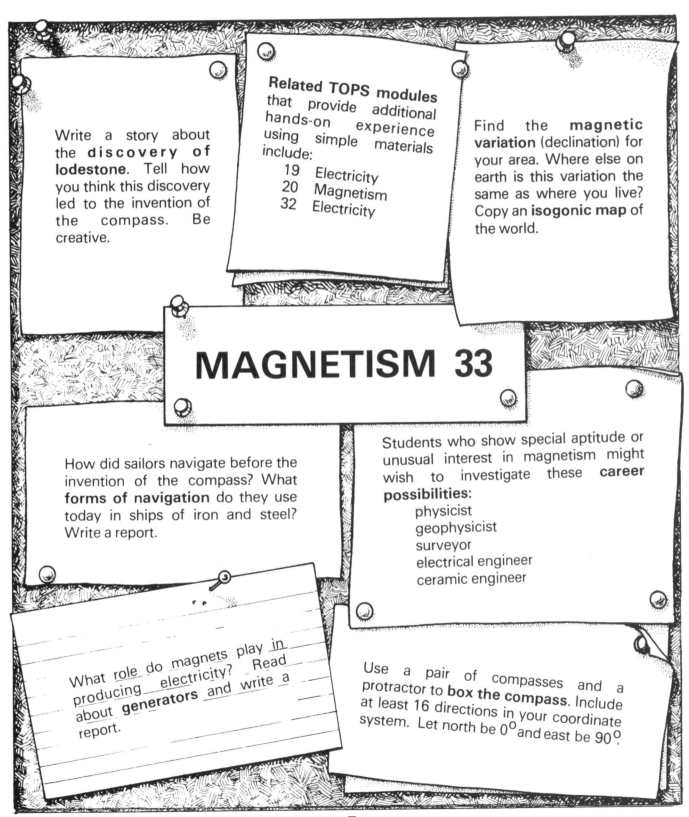

Write a story about the **discovery of lodestone**. Tell how you think this discovery led to the invention of the compass. Be creative.

Related TOPS modules that provide additional hands-on experience using simple materials include:

19 Electricity
20 Magnetism
32 Electricity

Find the **magnetic variation** (declination) for your area. Where else on earth is this variation the same as where you live? Copy an **isogonic map** of the world.

MAGNETISM 33

How did sailors navigate before the invention of the compass? What **forms of navigation** do they use today in ships of iron and steel? Write a report.

Students who show special aptitude or unusual interest in magnetism might wish to investigate these **career possibilities:**
physicist
geophysicist
surveyor
electrical engineer
ceramic engineer

What role do magnets play in producing electricity? Read about **generators** and write a report.

Use a pair of compasses and a protractor to **box the compass**. Include at least 16 directions in your coordinate system. Let north be $0°$ and east be $90°$.

Review / Test Questions

Photocopy these test questions. Cut out those you wish to use, and tape them onto white paper. Include questions of your own design, as well. Crowd them all onto a single page for students to answer on their own papers, or leave space for student responses after each question, as you wish. Duplicate a class set, and your custom-made test is ready to use. Use leftover questions as a class review in preparation for the final exam.

activity 1
The owner of a metal shop wishes to recycle the copper and iron shavings that are swept from the floor each day. How should she separate this mixture of metals?

activity 2
Name each pole (**N** or **S**) on the lines provided: pole **B** attracts pole **Y**. When magnet **XY** hangs from a thread, pole **Y** points to Earth's north.

activity 3
Three pins, **X**, **Y**, and **Z**, are magnetized to attract and repel like these:

a. Given this information, write whether these pin combinations also attract or repel:

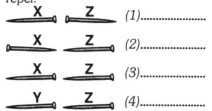

(1)......................

(2)......................

(3)......................

(4)......................

(Hint: You might begin by labeling the ends of all pins with circles and squares.)

b. If the **X** pinhead points *south*...
the **Y** pinhead must point ....
the **Z** pinhead must point ....

activity 4
Three magnets hang so they almost "grab" each other.

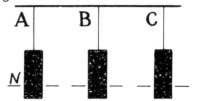

a. One pole is marked north (**N**). Label the other 5 poles.
b. Could you turn magnet **C** by turning magnet **A**? Explain.

activity 5
A compass needle is a tiny magnet that lines up with the earth's magnetic field to give directions. As a compass maker, would you make your compass cases from iron or aluminum? Explain.

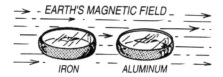

activity 6
What does this graph tell you about a magnet?

activity 7
The arrows represent atoms with magnetic poles. The circles represent atoms without poles. Identify each bar as a strong magnet, weak magnet, demagnetized material, or nonmagnetic material.

a. ....

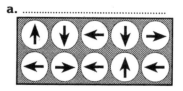

b. ....

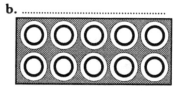

c. ....

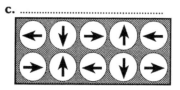

d. ....

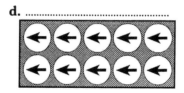

activity 8
On a backpacking trip, you accidently step on your compass. All you can save is the needle. Using only items you are likely to have with you or find around you, how could you fashion a new compass from the old needle?

activity 9
Make a small dot in the center of a clean sheet of paper, then tape the paper to your table. Use your ruler and hairline compass to solve this letter puzzle:

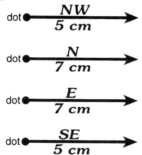

activity 10
Imagine that ants are walking in a straight line along the top edge of your blackboard from right to left.

a. Use your compass to find their direction of travel.
b. Tell how you did this.

Answers

activity 1

Pass a magnet through the pile of metal shavings. The iron will stick to the magnet, leaving the copper behind.

activity 2

Y is north by definition.
B is south because it attracts **Y**.
X is south because it's opposite **Y**.
A is north because it's opposite **B**.

activity 3

a. *(1)* attracts
   (2) attracts
   (3) repels
   (4) repels
b. ... pin **Y** must point **SOUTH**.
   ... pin **Z** must point **NORTH**.

activity 4

a.

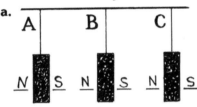

b. Yes, by a chain reaction between the magnets: **A** turns **B**, which turns **C**.

activity 5

Make it from aluminum. The earth's magnetic field passes through aluminum relatively undisturbed, but not through iron. If the cases were made from iron, the earth's field would be bent, causing the needles to stray from magnetic north.

activity 6

The attractive force of a magnet weakens as a magnetic object is placed farther and farther away. *(According to Coulomb's law, this force decreases inversely with the square of the distance between.)*

activity 7

a. weak magnet
b. nonmagnetic material
c. demagnetized material
d. strong magnet

activity 8

If you have a Band-Aid, you can suspend the needle from a strand of hair and tape it inside any non-iron container, say a styrofoam or aluminum cup. Or poke the needle through a bit of cork or styrofoam and float it on some calm water. Even a mud puddle would do.

activity 9

These compass directions form the letter **K**.

activity 10

a. The ants are walking _____.
b. First I rotated my compass so the printed needle and magnetic needle both lined up pointing north. Then I imagined how the ant would cross the center of my compass walking parallel to the blackboard.

H

Review / Test Questions, continued

activity 11

Finish this picture. Use long smooth lines to represent the lines of force in the magnetic field.

activity 12

Label the other 5 poles not shown.

activity 13

Finish this picture. Draw lines of force to show the shape of the interacting magnetic fields.

activity 14

Junk yard cranes use giant electromagnets to pick up old car bodies.

a. If you pulled this electromagnet apart, what would you expect to find inside?

b. As a crane operator, how would you pick up a car body and let it down again?

c. Would this crane work if you used a giant permanent magnet in place of an electromagnet? Why?

activity 15

This hat-pins compass points backwards.

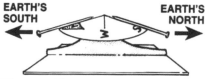

How would you remagnetize the pins so they point correctly?

activity 16

Tell how you would put together a compass, a dry cell and an electromagnet to make the compass needle spin like a motor. Draw a diagram.

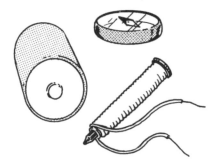

activity 17

Which invention came first, the electromagnet or the telegraph? Explain your reasoning.

activity 18

The hammer rings the bell by moving rapidly back and forth.

How might an electromagnet and a spring work together to make the hammer move?

activity 19

Will your on-off motor work without permanent magnets underneath the loop? Explain.

activity 20

Given a paper towel and two strong magnets, how could you wipe the algae off the inside of a piranha tank without putting your hands in the dangerous waters?

Answers, continued

activity 11

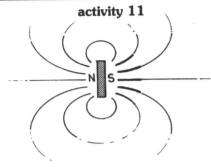

activity 12

activity 13

activity 14

a. You would find some kind of metal core wound with lots of insulated wire.

b. By controlling a switch that connects the wire coil to electricity: turn the switch on to pick up the car and turn it off to release the car.

c. No. If the magnet were permanent, you would not be able to turn off the magnetism to release the car.

activity 15

For the compass to point correctly, the pin head labeled *N* must be touched to *south* on a permanent magnet. In like manner, the pin head labeled *S* must be touched to *north* on a permanent magnet. *(This is because the earth's magnetic poles lie near oppositely named geographic poles. See activity 2, note 5-6.)*

activity 16

Put the compass near the electromagnet, and touch the wires leading from this electromagnet to the dry cell as shown. To make the needle spin, briefly turn the electromagnet on and off, once for every revolution of the needle.

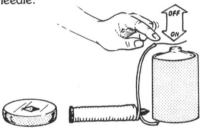

activity 17

The telegraph incorporates the electromagnet as one of its principle parts. The electromagnet, therefore, would have to exist before the telegraph was invented.

activity 18

An electromagnet and spring might work together, pulling the arm of the hammer first one way and then the other. Every time the spring pulls the hammer arm back, this motion completes the circuit that turns on the electromagnet. This attracts the hammer arm and stretches the spring. But this forward motion immediately breaks the circuit and turns the electromagnet off, allowing the spring to pull the hammer arm back again. This cycle repeats over and over to ring the bell.

activity 19

No. Without permanent magnets nearby, the coil would be neither attracted nor repelled when electricity flowed through it. So the coil would not turn. (Here we're assuming the earth's magnetic field is too weak to matter.)

activity 20

Wrap the paper towel around one of the strong magnets. Then place it just inside the top rim of the glass, opposite the other magnet on the outside. Because the magnets are strong, they will stay together, with the glass of the piranha tank in between. So you can wipe off the algae on the inside by moving only the outside magnet.

Long-Range Objectives

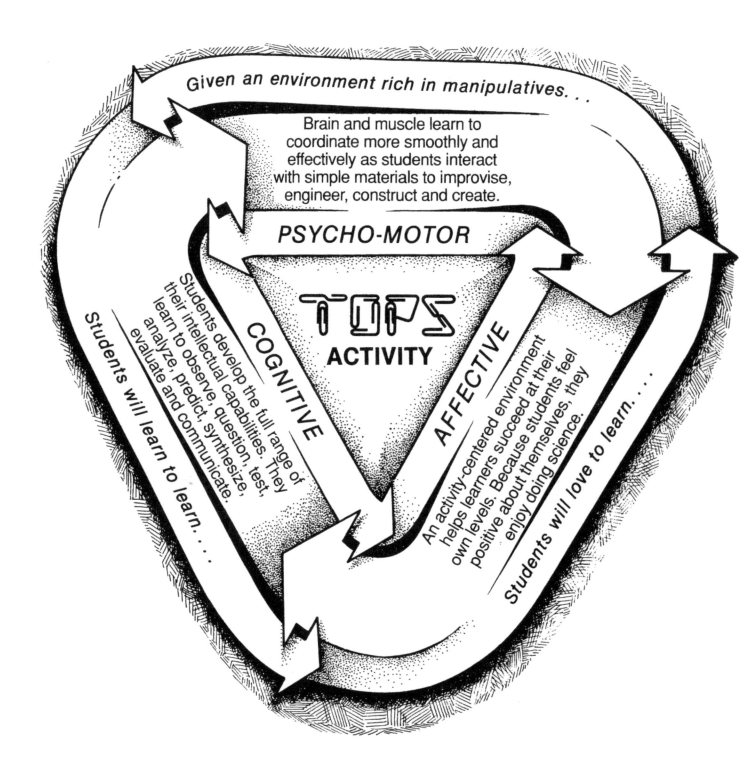

Given an environment rich in manipulatives. . .

Brain and muscle learn to coordinate more smoothly and effectively as students interact with simple materials to improvise, engineer, construct and create.

PSYCHO-MOTOR

TOPS ACTIVITY

COGNITIVE

Students develop the full range of their intellectual capabilities. They learn to observe, question, test, analyze, predict, synthesize, evaluate and communicate.

AFFECTIVE

An activity-centered environment helps learners succeed at their own levels. Because students feel positive about themselves, they enjoy doing science.

Students will learn to learn. . . .

Students will love to learn. . . .

K

ACTIVITIES
AND
LESSON NOTES
1-20

☞ As you duplicate and distribute these activity pages, **please observe our copyright restrictions** at the front of this book. Our basic rule is: **One book, one teacher.**

☞ TOPS is a small, not-for-profit educational corporation, dedicated to making great science accessible to students everywhere. Our only income is from the sale of these inexpensive modules. If you would like to help spread the word that TOPS is tops, please request multiple copies of our **free TOPS Ideas catalog** to pass on to other educators or student teachers. These offer a variety of sample lessons, plus an order form for your colleagues to purchase their own TOPS modules. Thanks!

IS IT MAGNETIC?

1 Fill in each table. List things *attracted* by a magnet on the *left* and things *not attracted* on the *right*.

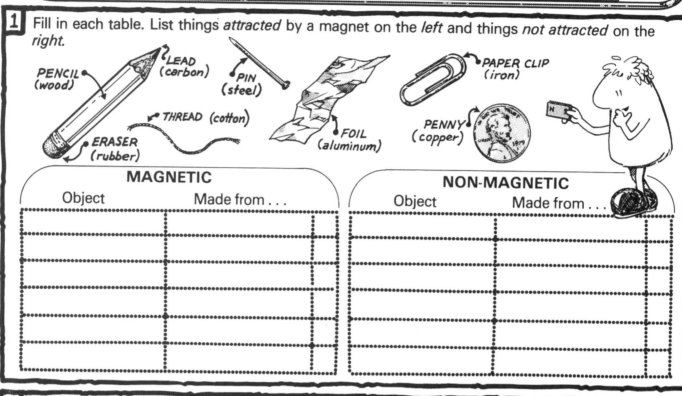

PENCIL (wood)
LEAD (carbon)
PIN (steel)
THREAD (cotton)
ERASER (rubber)
FOIL (aluminum)
PAPER CLIP (iron)
PENNY (copper)

MAGNETIC		NON-MAGNETIC	
Object	Made from . . .	Object	Made from . . .

2 Check those boxes above where the object is made from metal.

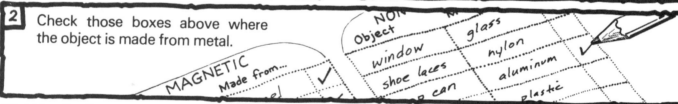

NON
Object
glass
window
nylon
shoe laces
aluminum
can
plastic
MAGNETIC
Made from...

3 Write the correct number in the correct space.

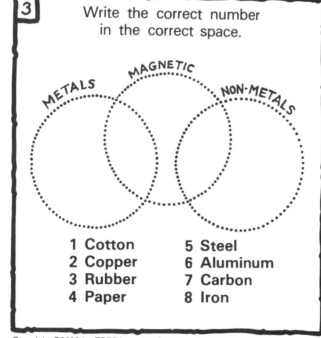

METALS   MAGNETIC   NON-METALS

1 Cotton   5 Steel
2 Copper   6 Aluminum
3 Rubber   7 Carbon
4 Paper   8 Iron

4 Circle true or false. Then give a reason for your answer.

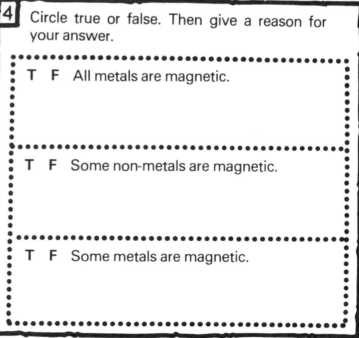

T F   All metals are magnetic.

T F   Some non-metals are magnetic.

T F   Some metals are magnetic.

TOPS LEARNING SYSTEMS

Objective

To recognize that only a few metals, like iron and steel, are magnetic, while most other metals and nonmetals are not.

Lesson Notes

To successfully teach this module you **must** have the right kind of magnets. Fortunately, the rectangular ceramic magnets you need are easy to get. You probably already have a few holding messages to your refrigerator door.

Buy ceramic magnets from science supply outlets, or electronics stores like Radio Shack, or from our TOPS catalog. These come in a variety of shapes and sizes. Some may have a hole in the center, others may be solid. A size about as large as the face of a postage stamp, or even a bit smaller, will work just fine.

ACTUAL SIZES:

Large traditional bar magnets or small circular ring magnets are **not** good substitutes.

Because these little magnets are relatively inexpensive, buy a bunch – at least 2 per student plus replacement extras – enough to accommodate your largest class. You can get by with less, of course, as few as 2 per activity group. But this will severely restrict individual involvement in the excitement of discovery learning.

1-2. Be prepared for lots of questions: What is it? How do you spell it? What's it made from? In general, any magnetic object found in your classroom will probably contain **iron**. Nickel and cobalt are the only other magnetic elements, but these are rare, and almost never found in concentrations large enough to be attracted to magnets.

Your students will also find magnetic things made from **steel**. Steel is an alloy of iron. It is usually mixed with carbon (sometimes with other metals) to improve strength and hardness. Iron is seldom encountered alone as a chemically pure element.

Whether you say a paper clip is made from iron or steel is a question of semantics: how much carbon should iron have before you call it steel? One way to decide is to consider the metal's hardness. Greater amounts of carbon are alloyed with iron to produce harder steel. A relatively soft **iron** paper clip might be distinguished from a hardened **steel** pin on this basis. But this distinction is of little consequence; allow your students to interchange the terms iron and steel as they please.

Other questions that may arise: A tin can is magnetic not because it is made from tin. Rather, it is made from tin-plated iron (or steel). **Ceramic** magnets are magnetic not because they are made from a special sort of clay. They are made from iron oxide. Nickel as an element is magnetic, although the U.S. nickel coin (a 25% nickel, 75% copper alloy) shows no visible attraction to a magnet. Canadian nickels, however, are magnetic.

3. For classes that are unfamiliar with the logic of sets, draw these Venn diagrams, on your blackboard.

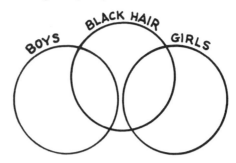

Have each student identify his or her position within the proper section of the proper circle. Where would you place a black wig? (In the center of the middle circle.) Where would you put a white wig? (Outside all three circles.) Where would you park a gray Studebaker?

Answers

1-2. Typical answers:

MAGNETIC			NONMAGNETIC	
Object:	Made from:		Object:	Made from:
pin	steel ✔		pencil	wood
paper clip	iron ✔		eraser	rubber
nail	iron ✔		pencil lead	carbon (graphite)
scissors	steel ✔		thread	cotton
staple	iron ✔		foil	aluminum ✔
coat hanger	iron ✔		penny	copper ✔

3.

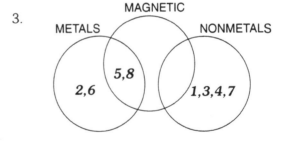

3. Ⓕ Copper and aluminum are metals, but nonmagnetic.

Ⓕ All nonmetals tested were nonmagnetic.

Ⓣ Iron and steel are examples of magnetic materials.

Materials

☐ Ceramic magnets. All activities in this module are specifically designed for small rectangular permanent magnets, often called ceramic magnets or sandwich magnets. (See opening paragraphs in the teaching notes above.) Purchase these on your local economy or order direct from TOPS. We will ship the quantity you specify and bill you our current catalog price plus shipping.

☐ A collection of objects to test for magnetic attraction. Be sure to include thread, aluminum foil and copper pennies; also paper clips, steel pins and other objects made from iron and steel.

NAME THAT POLE

1 Bring 2 magnets close together.

What happens when the magnetic poles ATTRACT?

What happens when the magnetic poles REPEL?

2 Cover both poles on each magnet with masking tape.

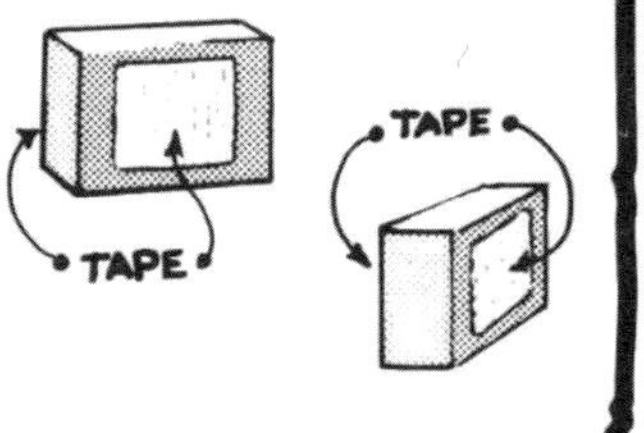

TAPE

TAPE

3 Draw large *circles* on two poles that repel.

CIRCLES

Draw large *squares* on the other sides.

SQUARES

4 Use your magnets to complete this table.

like poles	◯ to ◯	repel
unlike poles	◯ to ☐	
	☐ to ☐	
	☐ to ◯	

5 Tape a loop of thread to your table. Make it about as long as this paper . . .

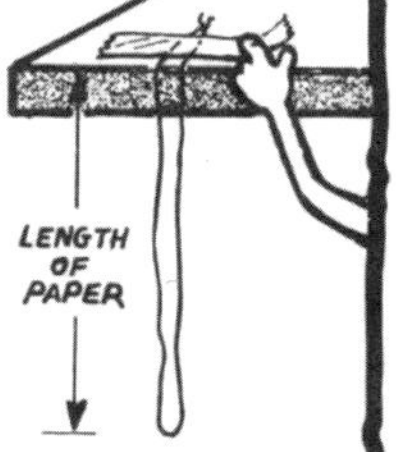

LENGTH OF PAPER

Then hook 2 paper clips into your loop . . .

Then hang one of your magnets between the paper clips. Allow it to come to rest.

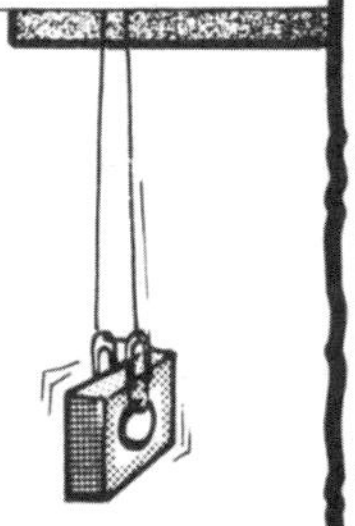

6 The ⎡circle or square?⎤ points north. Label it **"N"**.

The ⎡circle or square?⎤ points south. Label it **"S"**.

Label the north and south poles on your other magnet.

N or S?

7 Complete this check list to be sure you named the poles correctly.

If you hang either magnet from the loop:
☐ N must face earth's North
☐ S must face earth's South

When you bring your magnets together:
☐ S must repel S.
☐ N must repel N.
☐ S must attract N.

Show your magnets to your teacher.
☐ Teacher check.

TOPS LEARNING SYSTEMS

Objective

To identify and label the north and south poles on unmarked magnets by using the earth's magnetic field as a reference.

Lesson Notes

The poles on these flat ceramic magnets point outward from the broad flat face of each side. (Contrast this orientation with the poles on traditional bar magnets located at each narrow end.) In this activity your students will decide by experiment which of these poles is north and which is south.

5-6. *By definition*, the north-seeking pole is called north. On some magnets this will be identified with a square. On others it will be the circle.

Sometimes students will reasonably question why the north side of their magnet should point to earth's magnetic north. After all, like poles repel!

In reality, the earth's north magnetic pole is a *south* pole! And the earth's south magnetic pole is really *north*! This has long been a great source of confusion.

But the confusion would be considerably worse if map makers and geographers located *magnetic* north and *geographic* north half a world apart. So the convention stands. Encyclopedias and atlases show magnetic north next to geographic north, even though it really isn't. Earth is a mislabeled magnet.

6. To complete this step, your students will need to understand how their classroom is oriented relative to the magnetic north-south axis. If north is not already known to all, you'll need to make a sign. Write a large N on a sheet of paper and hang it on the wall so it indicates magnetic north relative to the center of your room. Or for an even better visual representation, hang or mount a meter stick showing the correct orientation where your whole class can see it:

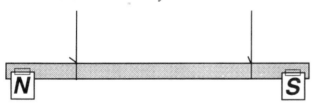

Magnetic north and geographic north don't quite coincide in most areas. This concept is treated as an extension in activity 10. For now, orient your sign to magnetic north, not true geographic north. If you are in doubt, consult a compass. (You can make one in activity 8.)

7. This step insures that each magnet is correctly labeled, If not, confusion will result in later activities.

Be sure you have labeled the poles on at least one magnet yourself so you know they are correct. You can do this by following the steps in this activity sheet. Or simply hang the magnet from some thread with the poles pointing out: It will always come to rest so the north pole faces north.

Label the poles on your checking magnet in large clear letters. When your students ask for a teacher check, hold your checking magnet in one hand and one of your student's

two magnets in the other. Ask, "Will these poles attract or repel?" Let your student predict, then test each hypothesis. This will confirm their labeling accuracy, solidify the concept of like and unlike poles, and provide good lesson closure.

After all the magnets are labeled and handed in, there is one final way to check for pole accuracy. Place all the magnets together in one long row. All north poles (and all south poles) will naturally face the same way. Any anomalies are easy to spot and should be relabeled.

Answers

1. When poles attract, they come together and remain attached.

When poles repel, they push each other away and remain apart.

4.

like poles	○ to ○	*repel*
unlike poles	○ to □	*attract*
like poles	□ to □	*repel*
unlike poles	□ to ○	*attract*

6. One of the boxes should be labeled N, the other S, in either order.

7. Students should have checked the first 5 boxes. You should mark the last box after your teacher check is complete. See note 7 for suggestions.

Materials

☐ Magnets. Use 2 per student or activity group.

☐ Masking tape.

☐ Scissors.

☐ Thread.

☐ Paper clips.

☐ A checking magnet with the poles properly labeled. See note 7.

☐ A sign to indicate magnetic north within your room. See note 6.

PIN MAGNETS

1 Cut out this small triangle. | Stick a pin through it like this: | Tape a tiny piece of tape to about 30 cm (1 foot) of thread.

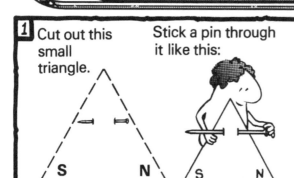

2 Hang it from your table. Be sure the pin is level and no iron is nearby.

LEVEL

3 Touch the pin head to *south* on a magnet so the triangle points correctly to earth's north and earth's south.

4 Tag 6 pin magnets with numbered paper punches.

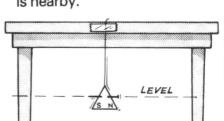

A STYRO-FOAM CUP PIN CUSHION

5 Bring each pin magnet near your pin triangle. Fill in the boxes with **N** or **S** to show how each pin is magnetized.

REMEMBER:

* LIKE POLES REPEL

* UNLIKE POLES ATTRACT!

6 Rest each pin magnet in a paper cradle that hangs from your pencil. Draw how each pin points in the boxes below.

1	S	N
2	S	N
3	S	N
4	S	N
5	S	N
6	S	N

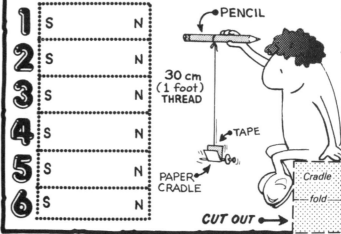

PENCIL

30 cm (1 foot) THREAD

TAPE

PAPER CRADLE

Cradle fold

CUT OUT →

7 Do your results in steps 5 and 6 agree? Explain.

8 Can you change a pin's poles back and forth with a permanent magnet? Explain.

TOPS LEARNING SYSTEMS

Objective

To experimentally determine the pole orientation of pin magnets.

Lesson Notes

1. The thread should be at least 30 cm long. Shorter threads are not as pliant, and if they get wound up, they can prevent the pin from achieving its preferred N-S axis.

3. If the pin head is already magnetized south (half the time it is), it will be repelled as you bring the south pole of the permanent magnet near. In this case you'll need to hold the triangle to prevent it from turning.

Once you touch the pin head to south on the magnet, you'll change its pole to north. It's hard to believe, but one touch is all it takes!

Touching the pin head to **south** on a magnet (thereby making it a north pole) so it will point to earth's **north** illustrates once again that earth's magnetic south pole is really located near geographic north. See activity 2, note 5-6.

4. Prepare pin magnets ahead of time, at least six per student or activity group. This is easy to do. Simply touch **half** the pin heads to north on a permanent magnet, and the other half to south. Then store them in a styrofoam cup labeled **PIN MAGNETS**. If you mix them well in this cup, the odds are good that all your students will draw pins from the cup that are magnetized both ways.

When your students ask for these pin magnets, it's a good time to collect their permanent magnets and not return them until step 8. The poles on pin magnets are reversible. You can change them just by a quick touch to a strong permanent magnet. If this inadvertently happens at any time during steps 5 and 6, there will not be good agreement in step 7. (Pin magnets won't change poles when touched to each other because they're not strong enough.)

Using only your fingers, it's difficult to stick a pin through a tiny paper punch; pin pricks are likely. A styrofoam cup used as a pin cushion makes this task much easier – and a lot safer.

6. Slide the paper punch tags all the way to the head of the pin. This will allow them to rest snugly in the paper cradle.

These pins will consistently point along a north-south axis *only* if the thread remains untwisted. For this reason, the cradle is attached to a long thread and a pencil. The pencil prevents students from rolling the thread between fingertips, subjecting the pin to a whirl of rotational force.

7. The tables in steps 5 and 6 must fully agree. If they don't, ask your students to recheck their work and locate errors.

8. Remember to return the permanent magnet that you removed in step 4.

Answers

5-6. Any numbered pin can be magnetized either N-S or S-N. But you should still check for relative consistency. The same pin cannot be labeled in one orientation in step 5 and the other orientation in step 6.

7. Yes. For every pin, the end that was magnetized north in step 5 pointed north in step 6.

8. Yes. Touch the pin head to south on a permanent magnet, and it points northward. Then touch the same pin head to north, and it points southward.

Materials

☐ Scissors.

☐ Steel straight pins. Be sure they are made from steel, not aluminum. Convert these into pin magnets with a 50:50 mix of magnetic orientation as detailed in note 4.

☐ Thread.

☐ Cellophane tape or masking tape.

☐ A magnet.

☐ Paper punches. Prepunch a good supply, or provide a paper punch for your students to use.

☐ Styrofoam cups.

INVISIBLE GEARS

1

Tie a loop of thread about as long as this paper . . .

Tape one of your magnets at one end of the loop . . .

TAPE

Then tape the loop to your table so the magnet hangs over the edge.

TAPE

THE POLES FACE SIDEWAYS, NOT UP AND DOWN

POLES ← →

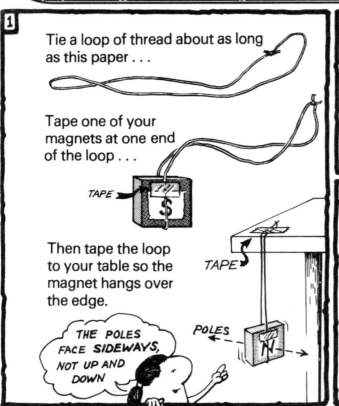

2

Hang another magnet level with the first. Keep them as close as you can without the 2 magnets grabbing each other.

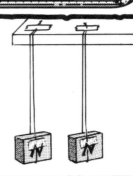

3

Rest one magnet on the table. Spin the other around and around until you wind up the string.

4

Allow both magnets to rest side by side. Then let them go.

LET THEM HANG WITHOUT SWINGING . . .

. . . THEN LET BOTH GO TOGETHER.

If they don't start spinning, give either magnet a gentle push.

Write your observations.

5

● As one magnet winds down, the other magnet ....

● As one magnet loses energy, the other magnet ....

● The magnets turn so that .... poles always face each other.

6

What is the invisible gear that links these 2 magnets?

Can this invisible gear:
  pass through this paper?

  pass through your hand?

TOPS LEARNING SYSTEMS

Objective

To observe interactions between rotating magnetic fields.

Lesson Notes

1. A loop as long as this paper requires two paper-lengths of thread, not one. Your students may overlook this and end up with only half the desired loop length. This may still work. Length is not critical so long as the loop is long enough to wind up properly in step 3.

Younger students may not know how to tie the loose ends to form a knot. They can simply tape the ends together on the table.

The magnet must hang in an upright position with the poles aimed out along the horizontal. It's best if you tape it to the thread along its top edge. If it still tilts, you can make minor corrections by pulling on one side of the loop or the other.

If your table legs are made from iron, remind students to hang their magnets near the center of the table, as far from the iron as possible. If the table top itself contains iron, make the loop long enough so the attraction between the magnet and the table top is reduced to a minimum.

2. The distance between the magnets is critical. It they are too close, the magnets won't be able to spin without locking together. If they are too far apart, the magnetic linkage effect in step 4 won't be as dramatic.

3. The fastest way to wind the magnet on its string is to give it a quick twist of the wrist or a sharp snap of the finger, then let it spin freely. When it comes to rest, spin it again in the same direction.

4. The secret here is to cradle both magnets between your fingers so they hang almost motionless side by side. Then let them go. If the one magnet is wound tightly enough, both will begin to spin on their own. If not, give one of them a gentle nudge.

Your students may not successfully observe the magnetic linkage right away. Encourage them to fine-tune the system, adjusting the spacing between the magnets until they are successful. Such trial-and-error tinkering is at the heart of scientific inquiry.

Answers

4. The magnets appear to be linked together. As one turns/ slows down/speeds up/stops/starts, so does the other.

5. *As one magnet winds down, the other* winds up.
   As one magnet loses energy, the other gains energy.
   The magnets turn so unlike *poles always face each other.*

6. Magnetic attraction is the "invisible gear."
   ...*pass through this paper?* Yes.
   ...*pass through your hand?* Yes.

NOTE: Some students may correctly observe that the interaction of the magnets through the paper is stronger than the interaction of the magnets through their hand. Then they will wrongly conclude that the hand somehow blocks the force of the magnets. The interaction is weaker only because the hand is thicker than the paper, keeping the magnets farther apart. Distance alone weakens the strength of the interacting magnets. This idea is explored in activity 6.

Materials

☐ Thread.

☐ Scissors.

☐ Magnets. Use 2 per student or activity group.

☐ Cellophane tape or masking tape.

UP IN THE AIR

1 Tape a magnet to the bottom of a cup so it hangs over the edge like a diving board.

2 Tie a paper clip to some thread. Touch it to the bottom of the magnet so the thread hangs down.

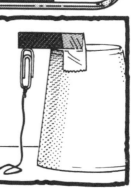

3 Tape the thread to the table, but leave the end free.

LEAVE END LOOSE

4 Now pull the thread so the paper clip hangs "up in the air" away from the magnet. Keep the space between as wide as you can.

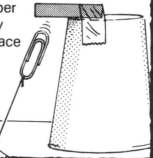

5 Pass each material below through the magnetic field that holds the paper clip.

Write your observations.

index card

pin

paper clip

aluminum foil

6 Predict if a penny will disturb the magnetic field. Give a reason for your answer.

Test your prediction.   ☐ correct   ☐ wrong

7 Predict if a washer will disturb the magnetic field. Give a reason for your answer.

Test your prediction.   ☐ correct   ☐ wrong

TOPS LEARNING SYSTEMS

Objective

To appreciate that a magnetic field can pass through solid objects, as long as they are not magnetic.

Lesson Notes

1. If the inverted cup becomes too top-heavy from the weight of the magnet above, you can stabilize it by taping the opposite side to the table.

4. You can also hang the paper clip in the air by sliding the cup slightly away.

5. The force that surrounds a magnet is known as a *magnetic field*. This field will pass right through nonmagnetic materials, like the index card and aluminum foil, with no visible effect on the paper clip. But the field will be significantly altered by magnetic materials, such as the pin and paper clip. In the presence of a field, these objects temporarily become magnets themselves, and thereby interact with the permanent magnet above and the suspended paper clip below. For a more detailed examination of interacting magnetic fields, see activities 12 and 13.

If anyone claims that the index card or aluminum foil caused the paper clip to wiggle or fall, they haven't exercised enough care in passing these test objects through the field. Younger students, in particular, may not be coordinated enough to avoid bumping the suspended paper clip.

6-7. The important generalization here is that only magnetic objects interact with the field and cause the paper clip to wiggle or fall. Thus the iron washer disturbs the field, while the copper penny does not.

NOTE: This book's classification of substances as *magnetic* or *nonmagnetic*, while suitable for an elementary understanding of magnetism, should nevertheless be recognized as a simplification. Nonmagnetic substances (implicitly defined in this module as materials that are not visibly attracted to a magnet) do, in fact, interact with magnets, although very weakly.

Whenever electrons move, they produce an associated magnetic field (see activity 14). In a sense, then, all matter is magnetic, because everything (except the most fundamental particles) contains electrons.

Most substances are *diamagnetic*. Their electrons weakly oppose external magnetic fields at either pole. A few substances, because of unpaired electrons, are *paramagnetic*. They weakly increase external magnetic fields at either pole. Only three elements, iron, nickel and cobalt, are *ferromagnetic*. Their electron configuration enables these elements to remain permanently magnetized (see activity 7).

In this book, ferromagnetic substances are simply called magnetic. The other forms of weak magnetism, being too subtle to observe directly and too complicated for elementary study, are not considered.

Answers

5. *index card:* The paper clip is not disturbed.

pin: The paper clip wobbles, and sometimes falls.

paper clip: The paper clip moves aside and falls,

aluminum foil: The paper clip is not disturbed.

6. No. Nonmagnetic materials (the index card and foil) did not disturb the field. Since a penny is nonmagnetic, it shouldn't, either.

7. Yes. Magnetic objects made from iron (the pin and paper clip) disturbed the field. So should an iron washer.

Materials

☐ Cellophane tape or masking tape.

☐ A magnet.

☐ A plastic cup, glass beaker or jar. You may wish to purchase the short, wide cocktail cups sold in supermarkets - at least 2 per student. These make excellent compass containers that your students can take home. See activity 8.

STRENGTH OF A MAGNET

1 Clamp a magnet in a clothespin and tape it to the bottom of a cup. Pull apart a paper clip (just a little) and hang it underneath.

TAPE BOTTOM WING

PULL OUT TO MAKE A HOOK

2 Add paper clips. Write the total number (support clip included) that the magnet can hold next to the "0" in the table below.

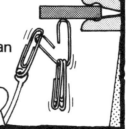

3 Cut a piece of masking tape the size of a small postage stamp and stick it underneath the magnet.

Now that the paper clips can't get quite as close to the magnet, how many can you hang?

WRITE YOUR ANSWER IN THIS TABLE.

4 Continue adding layers of tape. Add paper clips to complete the table.

STACK THE LAYERS NEATLY

5 Make a graph. Draw the best smooth curve you can through the points.

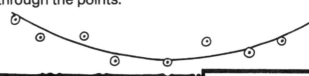

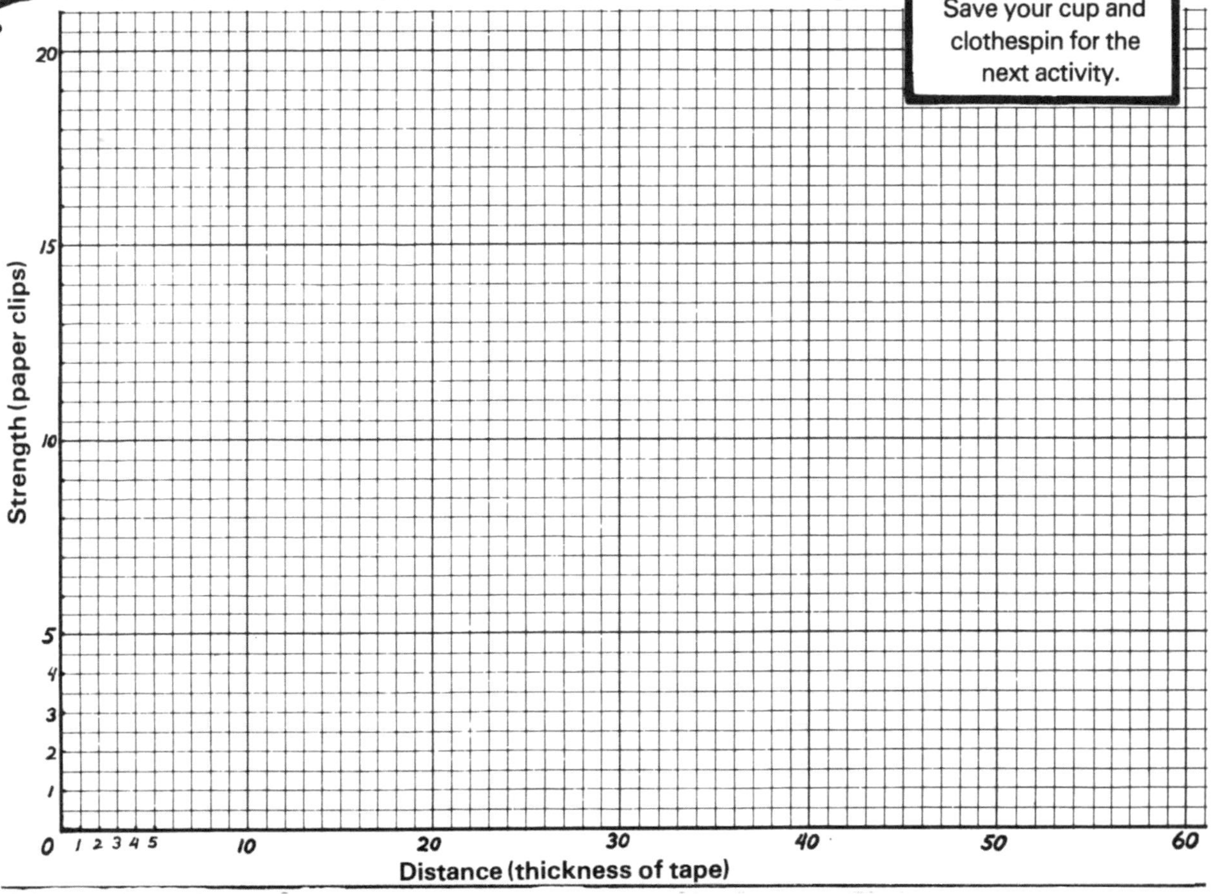

Save your cup and clothespin for the next activity.

DISTANCE (LAYERS OF TAPE)	STRENGTH (PAPER CLIPS)
0	
1	
2	
4	
8	
12	
20	
30	
40	
60	

Strength (paper clips)

20

15

10

5
4
3
2
1

0 1 2 3 4 5   10   20   30   40   50   60

Distance (thickness of tape)

TOPS LEARNING SYSTEMS

Objective

To graph how the strength of a magnet varies with increased distance from the magnet.

Lesson Notes

In this activity your students will measure the strength of a magnet by how many **paper clips** it supports at its pole. And they will measure distance in **tape thicknesses** (layers of tape) covering the pole.

Allow plenty of time to experiment and record data, perhaps an hour. If your students need to continue their experiments-in-progress the next day, be sure they use the same magnets.

1-2. Notice that only the bottom half of the clothespin is taped to the inverted cup. The top half remains free so you can easily remove the magnet from its jaws.

The support paper clip is bent just far enough out to allow paper clips to slip through the gap in the middle. Its shape prevents the added clips from scattering when they fall to the table. And they do fall – again and again – as students try to add just one more clip to the cluster under the magnet.

There is already one piece of tape stuck to the surface of the magnet, the one used to label the pole. Your students can think of this as an integral part of the magnet itself, so that magnet and tape form the zero-layer starting point. Or they can remove the label if they wish to truly begin with "zero layers," and put it back after they finish this experiment.

3. Be sure your students understand why they add masking tape. Activity 5 has already established that a magnetic field will pass through nonmagnetic substances (like tape) with no effect. They are not **blocking** the magnetic field by adding more tape. Rather, each piece of tape removes the paper clip from the surface of the magnet by one more small increment of **distance**, equal to the thickness of the tape.

4. Care must be exercised when adding layers of tape so the paper clips and magnet will always be separated by the correct distance. If the tape is not cut to uniform lengths and added evenly, a rounded mound will form instead of a flat, uniform surface.

If the tape is cut too long, so that pieces extend under the jaws of the clothespin, the magnet tends to tilt.

Some may not appreciate that the number of tape layers in the data table represents an **accumulated** total. They should not add, for example, 12 layers, then another 20 layers, then 30. Rather, they must add 4 new layers to make 12, 8 new layers to make 20, 10 new layers to make 30 and so on.

Sandra Ciolino of Hyde Park School in Cincinnati, Ohio, makes the following suggestion. Before the lesson even starts, ask students to cut 60 postage-sized pieces of masking tape and stick them lightly to a smooth manila folder. Number these piece from 1 to 60. This makes it easier to keep track of the total number of tape layers, since each numbered piece is used in consecutive order.

5. Even if errors do occur in collecting data, meaningful graphs are still possible. Your students should understand that they should **not** connect every data point on the graph with a line. Instead they should draw the best possible smooth curve to fit the overall trend suggested by the scattered points.

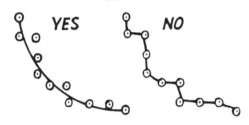

Answers

3-5. Typical answer: (Graph is turned sideways.)

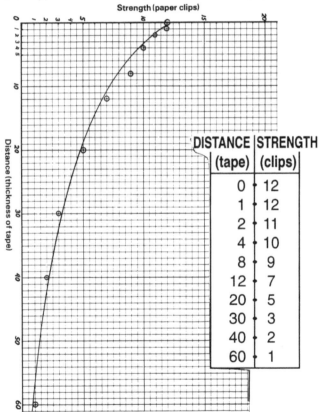

DISTANCE	STRENGTH
(tape)	(clips)
0	12
1	12
2	11
4	10
8	9
12	7
20	5
30	3
40	2
60	1

Materials

☐ A magnet.

☐ A clothespin.

☐ Masking tape. Distribute one roll between activity groups by peeling off long strips. See also note 4.

☐ A plastic cup, glass beaker or jar.

☐ Paper clips of uniform size and weight.

☐ Scissors. Supply at least one pair per activity group.

MAGNET MODELS

1 Compare the strength (in paper clips) of a pair of magnets that:

ATTRACT
(line up)

REPEL
(cancel out)

How many clips?

How many clips?

2 Get 8 pennies with arrows. Line up 4 of the arrows in these circles to represent "iron atoms" in one of your magnets.

ARROWS POINT NORTH!

POLE

POLE

Label the poles of this "magnet." Then remove the pennies and draw the arrows.

3 Arrange 8 iron atoms to represent a pair of magnets that **ATTRACT**. Draw the arrows.

4 Arrange these iron atoms to represent magnets that **REPEL**. Draw the arrows.

5 The 8 "atom" pennies have two sides.

ARROWS show atoms **WITH** magnetic poles.

CIRCLES show atoms **WITHOUT** magnetic poles.

Get the **ATOMS** page. Fill it in.

6 Think about the models you have just drawn to answer these questions.

a. If you cut a strong magnet in 2 parts, is each half still a magnet? Why?

b. Why are some magnets stronger than others?

c. Can you magnetize copper? Why?

d. How is a nonmagnetic material different from a demagnetized material?

TOPS LEARNING SYSTEMS

Objective

To develop a simple interpretive model to explain differences between strong magnets, weak magnets, magnetic materials, and nonmagnetic materials.

Lesson Notes

2. Each activity group will use 8 pennies. Draw arrows on the heads side, circles on the tails side. Make these as bold and visible as possible with a black permanent marker.

Students should place a penny in each circle, arrows showing, and turn them so all 4 arrows point either up or down. (The arrows should *not* point to either side. In this step they are modeling magnets with poles located on their faces.) Students should then label the resulting poles, remove the pennies, and draw arrows.

3-4. These steps recall the face-to-face magnet configurations in step 1, explaining why attracting magnets hold many more paper clips than repelling magnets. Student may question why cancelling "atoms" in the repelling magnets should support any paper clips at all. While the overall net polarity of canceling magnets is zero, there is still a nonzero local attraction caused by paper clips approaching closer to one magnet than the other.

5. These pennies represent atoms – the smallest particles of an element that carry the identity of that element. Thinking of iron atoms as tiny magnets is both useful and predictive on the elementary level. As students progress, they will refine and expand this simplistic notion to include more sophisticated ideas: Magnetic fields are generated by electrons with unpaired spin. These fields align among huge numbers of atoms to form domains, pockets of magnetism that grow and shrink as iron becomes magnetized and demagnetized.

One final thought about models. They represent reality in ways that help us understand how things work. We must always keep in mind that models are not the real thing. Billions upon billions of real atoms make up a real magnet, not the few arrows and circles pictured here.

As they work on the Atoms supplementary page, some students may not notice the distinction between *demagnetized* and *nonmagnetic* materials. Questions 8c and 8d will prompt them to take a more careful look.

Answers

1. Attracting magnets support perhaps 20 or 30 clips. Canceling magnets support as few as 2 to 6.

2. a magnet:

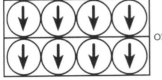

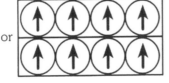

3. two magnets that ATTRACT:

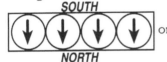

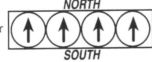

4. two magnets that REPEL:

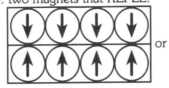
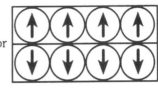

5. ATOMS supplementary page:

a. nonmagnetic aluminum

b. demagnetized iron

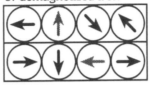

(All problems except a, c, d, and g have many possible answers.)

c. nonmagnetic copper

d. strongly magnetized iron

e. demagnetized iron

f. weakly magnetized iron

g. strong magnet   h. medium magnet   i. weak magnet

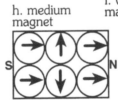
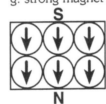
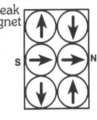

6a. Yes. Each half would still have its atoms magnetically aligned.

b. The atoms in stronger magnets have greater magnetic alignment than the atoms in weaker magnets.

c. No. Copper atoms have no magnetic poles, so no magnetic alignment is possible.

d. Nonmagnetic materials are made of atoms that do not have magnetic poles. Demagnetized materials have atoms with poles, but they cancel each other out.

Materials

☐ The cup-clothespin-magnet assembly from activity 6.

☐ Paper clips of uniform size and weight.

☐ Pennies and a black permanent marker. Draw arrows and circles on pennies per note 2.

HAIRLINE COMPASS

1 Lay a straight pin across sticky tape. Stick on a tiny arch of hair and trim ends.

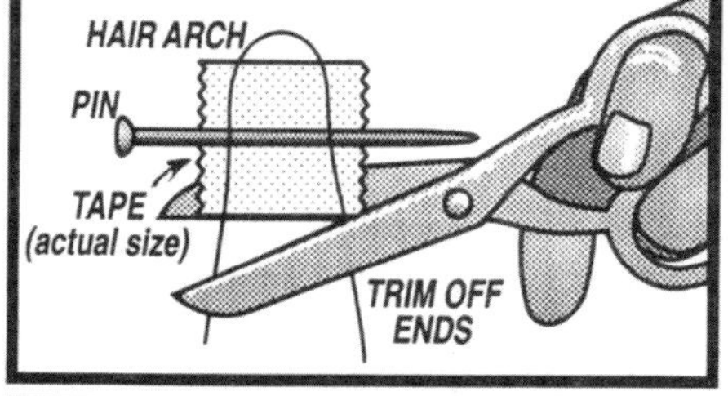

HAIR ARCH
PIN
TAPE (actual size)
TRIM OFF ENDS

2 Fold this tape around the pin. Adjust it to hang level from your pencil point.

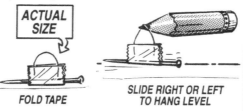

ACTUAL SIZE
FOLD TAPE
SLIDE RIGHT OR LEFT TO HANG LEVEL

3 Tape a long straight hair to the top of a clear plastic cup.

4 Thread the hair through the pin's arch, so the pin hangs near the bottom of the cup.

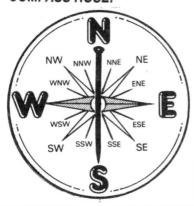

PASS IT THROUGH...   TAPE TO HANG NEAR BOTTOM

5 Make the pin a magnet: Touch the *head* of the pin to *south* on a permanent magnet.

COMPASS ROSE:

N
NW NNW NNE NE
WNW   ENE
W   E
WSW   ESE
SW SSW SSE SE
S

6 Cut out the compass rose. Fix it to the bottom of a second cup with rolled sticky tape.

7 Set the cups end to end, hairline compass on top.

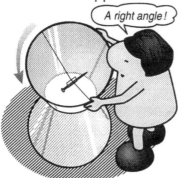

8 Turn the **top** cup until the pin hangs perpendicular to its hair support....

A right angle!

...holding this position, turn the **bottom** cup so the *compass* pin lines up with the *real* pin...

Keep that right angle!

...then tape the cups together in this special position.

Nice and neat...

TOPS LEARNING SYSTEMS

Objective

To build an accurate compass by suspending a magnetized pin from a strand of hair.

Lesson Notes

1-2. Be sure your students understand that the tape is illustrated **actual size**. There is a natural tendency to make the tape too large and the hair too long. Because they are working with small objects, younger, less coordinated students will require teacher assistance.

The hair should be stuck to the tape so it forms a wide, open arch, not a loop.

RIGHT   **WRONG**

It the pin does not hang level, adjust it by sliding it through the tape to a better point of balance.

3. Short, wide, plastic cocktail cups sold in supermarkets make excellent compass holders. You may substitute beakers or jars, as long as you can see clearly through the bottom to view the compass points taped underneath in step 8.

Even a soup bowl will work if you tape the compass directions inside. Cans or other containers made from iron will **not** work.

4. The pin should hang near, but not touch, the bottom of the cup.

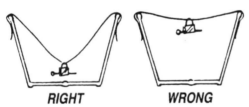

RIGHT   **WRONG**

6-7. A compass rose is called this for its round face and often ornate, petal-like layers of design. It is also sometimes called a compass card, especially when marked with up to 32 intermediate points and the 360 degrees of a circle.

Most desks contain some iron, if not in the surface, then in the legs. The bottom cup makes a convenient pedestal, raising the compass far enough above the desk surface that the iron should have no noticeable effect on the magnetized pin. If students can stay alert for nearby iron, they can tape the compass rose directly to the bottom of the top cup (either inside or outside) and omit the bottom support. A pile of books, or any other iron-free object, can always serve as a temporary pedestal.

Students may wonder why their compasses are so easily affected by nearby iron. The simplest explanation is that the pin, being a magnet, attracts iron. The desk can't move toward the pin, but the delicately suspended pin can move toward the iron. Another way to understand compass sensitivity is as a response to disturbances in the magnetic field that surrounds it. The needle faithfully lines up with lines of force that have been distorted or bent by iron interacting with earth's magnetic field. See extension in activity 13.

8. Unlike a conventional compass that pivots 360° on a pin, this overhead hairline suspension system allows the pin only about 120° of freedom. For this reason the compass directions must be taped to the cup so the pin **head** is able to point **north**.

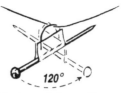

These directions assure that the pin has maximum freedom to swing about the northerly compass positions. Younger students may require extra supervision to be sure they position the compass coordinates properly.

Teacher Check

Ask your students to point out a specific direction, say east, using their compass. See that they first line up both pins (the one drawn below with the real one above). Then, without turning the compass farther, they can read east, or any other direction, directly from the compass coordinates.

If the hanging pin won't point north, magnetize it again by touching its head to south on a magnet. Then make sure no iron or steel objects are nearby. If the compass still won't work, check the hairline suspension system. The pin should have maximum freedom to turn when seeking north. If movement is restricted, untape the cups and repeat steps 7 and 8.

Materials

☐ Scissors.

☐ Cellophane tape.

☐ Straight long hair. Kinky or curly hair may not work. Thread is a poor substitute.

☐ A steel pin.

☐ Two plastic cups. See notes 3, and 6-7 above.

☐ A magnet.

LETTER PUZZLES

1 Make a small dot in the center of a clean sheet of paper.

2 Tape the paper to a level surface, away from magnets or other iron objects.

3 Cut out the ruler below.
Use this ruler and your hairline compass to solve each letter puzzle.

The ruler gives *distance*, the compass gives *direction*.

4 **FIRST LETTER PUZZLE:**

dot● ―――― *NE* / *8 cm* ――→

dot● ―――― *S* / *8 cm* ――→

dot● ―――― *NW* / *8 cm* ――→

*DRAW **3** STRAIGHT LINES **FROM THE CENTER** DOT.*

AS YOU SOLVE EACH PUZZLE, WRITE THE LETTER IN THE BOX.

5 **SECOND LETTER PUZZLE:**

Start again with a clean sheet of paper taped to your table, and make a small dot in the center.

dot● ―― *N* / *8 cm* ―→ ―― *SE* / *12 cm* ―→

dot● ―― *W* / *12 cm* ―→ ―― *SE* / *12 cm* ―→

*START FROM THE DOT JUST **TWICE**.*

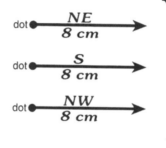

6 **THIRD LETTER PUZZLE:**

Start again with a clean sheet of paper taped to your table, and make a small dot in the center.

dot● ― *W* / *6 cm* →― *S* / *7 cm* →― *E* / *7 cm* →― *N* / *12 cm* →

dot● ― *NW* / *8 cm* →

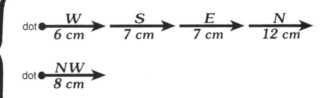

CENTIMETER RULER

| 0 | 1 | 2 | 3 | 4 | 5 | 6 | 7 | 8 | 9 | 10 | 11 | 12 | 13 | 14 | 15 |

TOPS LEARNING SYSTEMS

Objective

To practice plotting compass directions with a ruler and magnetic compass.

Lesson Notes

This book provides four levels of compass experience. As an introductory warm-up, try Extension (1) below. Then do activity 9 followed by activity 10. If you want your students to try a less structured, more advanced outdoor compass activity, try Extension (2).

4. These directions imply that you begin at the dot, then draw a straight line NE for 8 cm. Begin *at the dot* again and draw a straight line S for 8 cm, etc.

5. Here you begin at the dot and draw two lines end to end before *returning* to the dot a second time.

Extension (1)

PENCIL POINTING

Teacher-Student Activity: Write a compass direction on the blackboard, SE, for example. Your students should find SE on their hairline compasses, then point their pencils in that direction. Ask how all pencils now line up in relation to each other: they are all *parallel*. In special cases, these pencils might also line up end to end, but this is not a necessary condition for pointing in the same direction. The same is true for people. When walking SE, they might move in single file *or* side by side.

Repeat as necessary, using other compass directions.

Student-Student Activity: Divide your class into small activity groups of 2 or 3 students each. Someone in each group spins a pencil on the table. Each one in the group then uses the compass to decide how the pencil points, and writes down a specific direction on a separate paper. After all the answers are in, they compare their results for accuracy.

Students will debate among themselves who is right and who is wrong, learning much about compasses and their inherent accuracy in the process.

Repeat as necessary until the groups are generally able to agree about compass directions within a reasonable limit of accuracy. In general, answers should not vary by more than one compass mark in either direction (a maximum of 45°).

Extension (2)

TREASURE HUNT!

Everyone loves a treasure hunt. All you need is a large open area outside, two index cards, the hairline compasses from activity 8 and a small treasure to bury. You can make the treasure map ahead of time, or have your students make maps for each other to follow.

To make the map, first divide an index card into 8 map markers of equal size. If your entire class is using the same general outdoor area, students should identify their markers with their own initials.

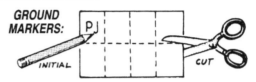

GROUND MARKERS:

Start from a familiar landmark like a tree or the school steps. Choose a compass bearing, then walk along this line a measured number of paces. Record this information on a full index card and place a map marker at your feet. Then choose a different direction and repeat the process. If you use up all 8 markers, your map will look something like this.

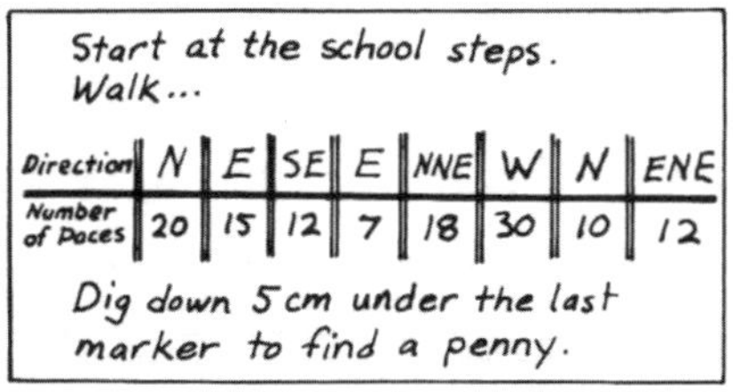

Start at the school steps.
Walk...

Direction	N	E	SE	E	NNE	W	N	ENE
Number of Paces	20	15	12	7	18	30	10	12

Dig down 5cm under the last marker to find a penny.

Give the map to another student to follow. Finding each marker may not be easy because both the map maker and the map follower must be accurate. Students will also need to define how far a pace should be, and learn to walk in measured units.

Answers

4. Y
5. M
6. R

Materials

☐ The hairline compass from activity 8.

☐ Tape and scissors.

WHICH WAY?

1 Notice that 4 corners in your room are marked with letters. Use your hairline compass to find each direction from the center of your desk.

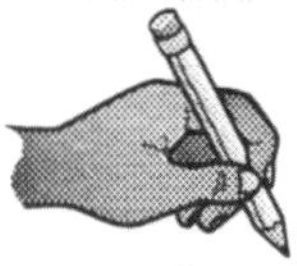

| WRITE EACH DIRECTION |

Corner A is .... of my desk.

Corner B is .... of my desk.

Corner C is .... of my desk.

Corner D is .... of my desk.

2 Draw a bird's-eye view of your room on a full sheet of paper:

✔ Measure distances with a meter stick.

✔ Draw walls, doors, windows and desks to scale. Be accurate!

✔ Label corners A, B, C and D.

3 Cut out this little "compass button." Tape it to your map over the drawing of *your desk*. Be sure to turn it in the correct direction.

YOUR DESK:

4 Do your directions in step 1 agree with the compass circle on your map? Give examples.

TOPS LEARNING SYSTEMS

Objective

To draw a birds-eye view of the classroom as accurately to scale as possible. To compare compass directions on this map to actual compass directions in the room.

Lesson Notes

1. Students should set their compasses in the middle of their desks and wait for the needle motion to stop. If the compass rose is then turned to its correct alignment with the suspended pin without lifting the assembly from the table, the needle will remain steady.

2. Require each student to work to maximum ability. For students of lower ability, a careful freehand drawing might suffice. To challenge higher skill levels, ask students to take room measurements and draw straight lines and right angles with an index card. To calculate scale, let y cm = x meters.

To begin, encourage all students to lightly sketch the shape of the room in approximate proportions, using all the space available on a sheet of paper. Many students will make tiny drawings if left to their own inclinations.

4. A piece of thread or string will assist students in making accurate map readings.

Extension

Earth's magnetic poles are near to (but do not coincide with) Earth's geographic poles. Therefore, from most places on the globe, there will be a directional variation between *magnetic* north and *geographic* north. This difference is called the **angle of variation**. (Surveyors call it the *angle of declination*.)

To illustrate this in concrete terms, draw a vertical line **representing** true (geographic) north on the blackboard. Hang a meter stick several meters in front of this line to **represent** magnetic north. Aim two pencils at these respective poles. Notice how the angle of variation changes east or west as you move about the room.

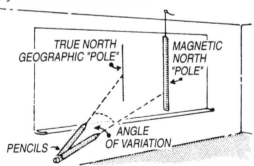

Make a large model compass by drawing bold coordinates on a full sheet of paper. Place a third pencil at the center to represent a compass needle.

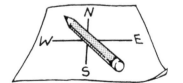

Stand somewhere in the classroom. First use your two pencils to decide which way the variation goes (toward east or west). Then align your model compass to the magnetic and geographic poles: Always aim the pencil (magnetic needle) at the magnetic pole, and the north compass coordinate at geographic north.

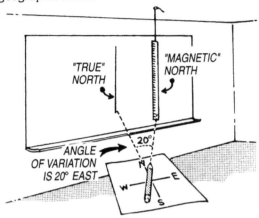

Find the magnetic variation for your geographical area by examining a local topographical map or searching the internet. If you find that your local variation is x° east, allow the magnetic needle to rest x° east of north as a correction; If your local variation is y° west, allow the magnetic needle to rest y° west of north as a correction.

Answers

1. Answers will vary depending on your choice of reference points and student desk positions relative to these points.

2. Room maps. Accept nothing less than each student's best effort.

3. Check that the compass button has true orientation. When it points north, is the map correctly aligned to the room?

4. Example: Using my compass, I determined that corner A is SSE from the center of my desk. When I run a piece of thread from the compass button to this same corner on my map, the thread runs only slightly east of SSE. This difference is within the bounds of acceptable experimental error... (and so on).

Materials

☐ Hang a sheet of paper in the 4 corners in your classroom. Label the corners A, B, C and D respectively.

☐ The hairline compass from activity 8.

☐ Scissors.

☐ Clear tape

☐ A meter stick, index card, and thread are optional.

MAP A MAGNETIC FIELD

1 Tape a short piece of thread to a pin. Make sure it hangs level.

2 Tape your magnet to the bottom half of this paper. Keep the poles pointing out as shown.

3 Touch the head of the pin to south.

4 One pin chain is drawn as an example. Draw 13 more pin chains, beginning each time at the points near the magnet.

FIRST, TEST HOW THE PIN HANGS...

...THEN DRAW ITS POSITION.

S N

TAPE MAGNET HERE

TAPE MAGNET ON THIS PAPER.

TOPS LEARNING SYSTEMS

Objective

To map the shape of a magnetic field that surrounds a magnet.

Lesson Notes

2. Students should tape a magnet directly to the lower half of their activity sheets in the space indicated. The poles **must point sideways**, not up and down.

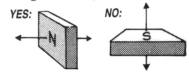

4. You can eliminate a lot of confusion by demonstrating this step to your whole class. Draw only one or two pin chains. Don't reveal the crab-like symmetry of the entire field. Allow your students the satisfaction of discovering this on their own.

During your demonstration, be sure to emphasize these points:

a. Begin at the points drawn near the magnet. Keep the pin on a **short "leash"** while it is close to the magnet, or it will be overcome by the strong pull and stick to the magnet.

b. Draw short, **half-length** pins, each time beginning where the previously drawn pin ends. Half-length pins give a better definition of the field, particularly in tight curves, than drawing full length pins.

c. As you map **farther** from the magnet, allow the pin a **longer "leash."** This will prevent the stiffness of the thread from overpowering the weakening magnetic field, and allow the pin to seek proper orientation.

d. It the thread gets twisted, exerting too much of its own influence on the pin, move the pin away from the magnet. Let it unwind, then bring it back into the magnetic field.

Some students, thinking the pin should always point straight out from the magnet, may complain that "it turns in the wrong direction." The pin, of course, simply responds to all the forces acting upon it. Student misconceptions are at fault, not the pin. Being objective (seeing things as they really **are**, not how we **think** they should be) requires a lifetime of wisdom. Assist this process by observing that it would be rather silly to blame the pin for misbehaving!

Because the pin **head** is touched to south, it is attracted to south and repelled away from north. If you think of the pins as arrows (pointing **head first**),

this correctly suggests a convention agreed upon among scientists: that the magnetic lines of force leave the north pole of a magnet and enter its south pole.

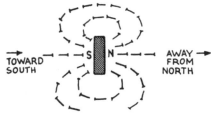

Starting points printed near the magnet suggest pin heads and pin points. These suggest that lines of force in a magnetic field are directional.

These particular starting points were selected to help your students achieve a clear, uncluttered map. They begin just far enough out to keep the magnet from "grabbing" the pin, and they are spaced evenly to give a fairly symmetrical pattern. Any other starting point, of course, will also work, since the magnet is surround by an infinite number of force lines.

Extension

Show how steel wool "dust" maps the magnet's field to produce a similar shape. See activity 12.

Answers

4. (Map is turned sideways.)

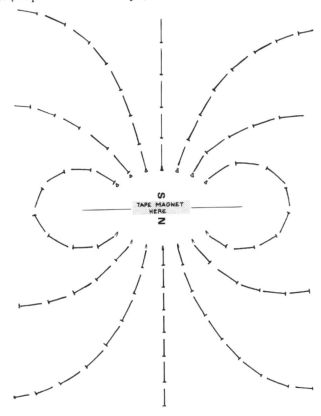

Materials

- ☐ Thread.
- ☐ Scissors.
- ☐ Tape.
- ☐ A steel pin.
- ☐ A magnet.

OPPOSITE FIELDS ATTRACT

1 Tape a magnet on each line below as shown. Be sure S faces N so both fields *ATTRACT*.

. . . then map the lines of force with a pin and thread.

FIRST TOUCH THE PIN HEAD TO SOUTH.

2 How do you think the fields from these magnets should look? Draw long *smooth* lines, not pins.

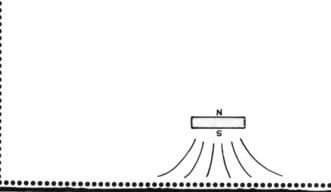

N
S

N
S

TAPE MAGNET HERE
N   S

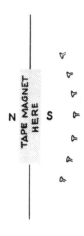

TAPE MAGNET HERE
N   S

TOPS LEARNING SYSTEMS

Objective

To map the shape of two interacting magnetic fields that attract.

Lesson Notes

1. Students should tape two magnets to their activity sheets so that south faces the marks representing pin heads, while north faces the pin points. Those who make a mistake here may draw lines of force that move in the wrong direction. Or worse, they may end up with repelling magnetic fields, the topic of activity 13.

Thinking of the pins as arrows, the lines of force move away from north and toward south, assuming a shape that somewhat resembles an onion.

If the magnets are correctly positioned but the lines of force still move in the wrong direction, the pin has been magnetized backwards. It should be remagnetized by touching the pin head to south as directed.

2. This question is asking for an educated guess based on the experience of step 1. Those who wonder whether they guessed right can make a pin map of the offset magnets on a separate sheet of paper. Or you can make a more detailed mapping with steel wool. See the extension activity that follows.

Extension

STEEL WOOL MAPPING

Another way to map a magnetic field is by using steel wool. Avoid the presoaped products sold in supermarkets. Look for plain steel wool in your local drug or hardware store. Choose a fine grade with fibers as thick as your hair.

To do the map, simply place one or more magnets on clean paper in any desired pattern. The poles, as usual, must point sideways. Then rub steel wool vigorously against itself so that small fibers settle over the entire paper. (Or you can sprinkle traditional iron filings over the magnets.)

While mapping this way is easy and provides great detail, it is messy and you risk slivers in your fingers. If you choose to make this extension a class activity instead of a teacher demonstration, consider supplying gloves. Afterward, the magnets should have their accumulated iron bristles removed. A quick trick is to press tape on the fibers, then peel it off.

Answers

1.

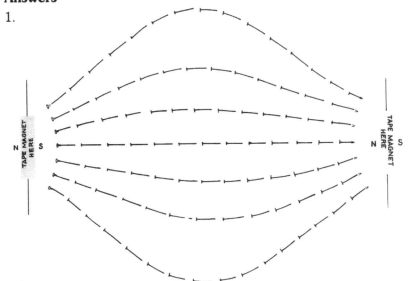

2.

Materials

☐ A steel pin suspended from a short piece of thread.

☐ Two magnets per student or activity group. If possible give every student the opportunity to individually map the attracting field here and the repelling field in activity 13. If you don't have enough magnets to go around, consider allowing part of your class to work ahead on activities 14 and 15.

LIKE FIELDS REPEL

1 *Predict* how you think the fields from these magnets should *repel*. Draw your best guess using long smooth lines of force.

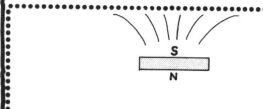

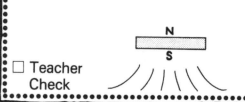

☐ Teacher
Check

2 Map the fields below to see if you are right. Be sure the magnets are placed so the poles *repel* each other:

TOUCH THE PIN **POINT** TO NORTH

NORTH **FACING** *NORTH*

TAPE MAGNET HERE

S   N

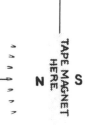

TAPE MAGNET HERE

N   S

TOPS LEARNING SYSTEMS

Objective

To map the shape of two interacting magnetic fields that repel.

Lesson Notes

1. Encourage thoughtful predictions. If the lines of force repel, they should in some manner avoid each other or push each other away. Accept any shape that in some way suggests repulsion.

2. Students should again tape two magnets to their activity sheets, this time with north facing north so the fields repel. When the pin is correctly magnetized (point touched to north and/or head touched to south), the lines of force will appear to leave each magnet, then veer off in a mutual repulsion pattern before reaching the center.

This open, diamond-shaped central area is where the dominance of one field ends and the opposite influence of the other field begins. The magnetized pin turns 180° as it moves through this area.

Notice that the starting points for the pin chains are bunched closer to the center of the magnet than with previous pin mappings. By directing more of the pin chains toward the center, they better penetrate the area of maximum repulsion. To close this diamond area even further, draw extra pin chains leading out from the center of each magnet.

The center of this diamond-shaped repulsion area cannot be meaningful mapped. Held over this area, the magnetized pin assumes a random orientation, responding unpredictably to the balance opposing fields, as well as twisting forces within the thread leash.

Steel wool mapping gives the most detailed definition of how magnetic fields repel as described in activity 12.

Extension

Encourage students who are intrigued by the patterns in interacting magnetic fields to try some independent investigations. Here are some possible configurations they might study.

Magnetic fields are also distorted in fascinating ways by placing iron or steel objects nearby. (And they are not distorted by nonmagnetic materials.) Make a few nails or washers available so curious kids can try them in various configurations that touch their magnets or rest close by. Students may also wish to experiment with a few nonmagnetic objects.

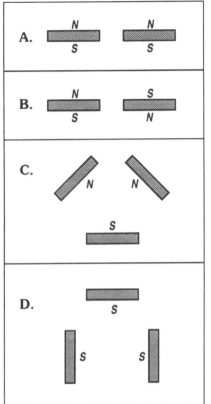

Answers

1.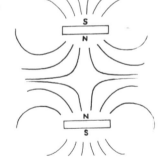

2.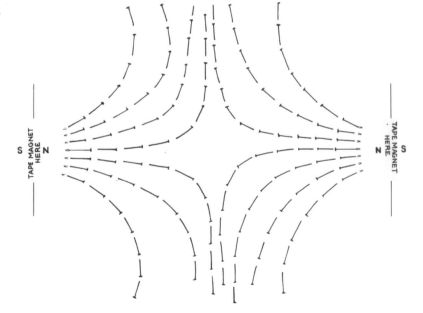

Materials

☐ A steel pin suspended from a short piece of thread.
☐ Two magnets per student or activity group.

BUILD AN ELECTROMAGNET

1 Begin at the point of the nail. Wind *insulated* wire, tightly and evenly, down to the head of the nail, then back to the point.

START→   ←FINISH

LEAVE LONG WIRE ENDS.

A   B   C

2 Peel off the insulation on each end. To do this *gently* cut around the outside with scissors, then pull the insulation off.

SLIDE OFF

NIP GENTLY

3 Clamp your electromagnet in a clothes-pin and tape one side to a dry cell.

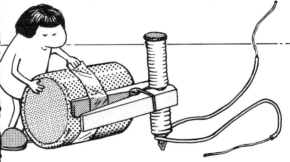

4 Put washers in circles A, B and C *above.* Use your electromagnet to move them to circles X, Y and Z *below.* It is not "legal" to drag or touch them with your fingers.

5 Tell how you picked up and released the washers with your electromagnet.

6 Is your electromagnet a temporary magnet or a permanent magnet? Explain.

~ **Save your electromagnet for later activities.** ~

X   Y   Z

TOPS LEARNING SYSTEMS

Objective

To learn how to construct and use an electromagnet. To appreciate that electromagnets are temporary, working only when electricity passes through the coil.

Lesson Notes

1. Cut the insulated wire into premeasured lengths. About 1½ meters (5 feet) of 24 gauge wire is suitable for medium sized 6½ cm (2½ inch) nails. If you use longer nails, you'll need extra wire. For shorter nails, use at least the same length of wire and wind it extra turns. This electromagnet will be used to build a telegraph in activity 17 and a buzzer in activity 18.

2. If the insulation is made from soft rubber, sharp fingernails might do the job.

3. Taping the electromagnet to the dry cell makes it easier to handle. This is especially important for younger, less coordinated students.

4. Unless they have had previous experience with wires and

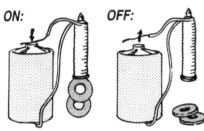

dry cells, many students will not know how to operate their electromagnets. Allow them to discover by trial and error how to turn them on and off.

Operating the electromagnet with both hands, it is neither necessary, nor a good idea, to tape either wire lead to the battery. The electromagnet uses a lot of energy. If taped and inadvertently left connected, the cell will quickly drain. Caution your students to use their cells no longer than necessary. They will use them again in later activities.

5. This question asks for an operational explanation: what you *do* to make the electromagnet work?

A deeper question is to ask *why* this electromagnet works? (Copper and aluminum wire are, after all, are both nonmagnetic materials.) Magnetism arises whenever charge moves. Electrons in motion create both electricity *and* magnetism, generating an associated electromagnetic field. When the electricity stops, so does the magnetic field. Students who wish to explore this relationship in more depth should try the extension that follows.

Extension
BUILD A GALVANOMETER

Whenever an electron moves, it has an associated magnetic field. As electricity moves around a coil, it creates a magnetic field through the coil. This field is strengthened by the presence of an iron core, such as a nail. But magnetic fields are also produced by hollow coils. To illustrate, try building this galvanometer (electricity tester).

Tape a loop of insulated wire to the cover of a book. Hang a *magnetized* pin from a short strand of hair and tape it to the top of the loop. Turn the book so the pin hangs *parallel* to the plane of the loop.

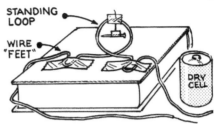

Now connect the loop momentarily to a cell. How does your galvanometer detect electricity? (The needle deflects to line up with the magnetic field produced by the moving electrons.) How can you make the needle jump the other way? (Reverse the cell.)

Answers

5. Touch the wire leads from your electromagnet to the ends of the dry cell (either direction works). This makes the nail magnetic, allowing you to pick up a washer. To release it again, simply disconnect one of the wire leads from the dry cell. The washer will drop immediately.

6. It's a temporary magnet. The magnetic field is maintained only while electricity flows through the coil. Break the circuit, and the nail looses most of its magnetism.

Materials

☐ A 6½ cm (2½ inch) iron nail. See note 1 above.

☐ About 1½ meters (5 feet) of 24 gauge insulated wire. See note 1 above.

☐ Wire cutters (optional). You can also cut the wire with an old pair of dull scissors, or bend it back and forth until it breaks.

☐ Scissors or a knife to strip off wire insulation. Prestrip the insulation for younger children.

☐ A size D dry cell. It should be fresh enough to last throughout 4 activities.

☐ A clothespin.

☐ Tape.

☐ Washers, from 3 to 9 per student or activity group.

HAT-PINS COMPASS

1 Fold an index card in half, then back again like an "M". Tape this "M" together forming "airplane wings."

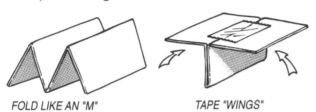

FOLD LIKE AN "M"   *TAPE "WINGS"*

2 Turn the card over so the middle stands up. Draw a "tower" leaving a flat space on top.

FLAT SPACE

3 Cut out the tower. Tape one pin to the side so it sticks up like a flag pole.

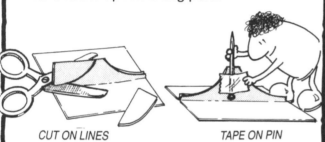

CUT ON LINES   *TAPE ON PIN*

4 Cut out the circle below. Now cut **in** along the dotted line. Overlap so both w's come together and tape.

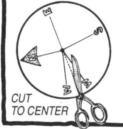

OVERLAP W'S

CUT TO CENTER   *TAPE*

5 Tape 2 pins along the N-S line so the heads point out.

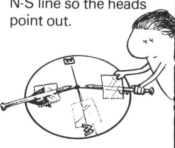

6 Touch each pin head to its opposite pole.

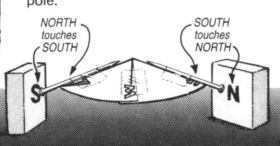

NORTH touches SOUTH   *SOUTH touches NORTH*

Carefully cut out this circle.

7 Balance these pins on the index card base. Compare this hat-pins compass to the hairline compass you made in activity 8.

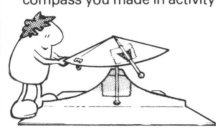

☐ Teacher Check

How are they the same? Different? Which do you like best?

(Continue answer on back if necessary.)

SAVE YOUR COMPASS FOR THE NEXT ACTIVITY.

TOPS LEARNING SYSTEMS

Objective

To build a compass by balancing two magnetized pins upon a third.

Lesson Notes

1. This index card must be folded squarely and evenly to form a flat, stable base for the compass.

4. The apex of this "hat" forms at the end of the center cut. It is important to stop cutting when you reach the center. If students go beyond, you can mend the over-cut with a bit of tape.

To view the overlapping W's, hold the hat up against strong light. When the two W's appear as one, the hat will form a broad, gently-sloping cone, steep enough to balance on the pin base in step 7, but shallow enough for the magnetized pins to point almost horizontally toward the earth's poles.

6. The pin taped to **north** on the compass must be touched to **south**. Likewise, the pin taped to south on the compass must be touched to north. Compasses that point the wrong way in step 7 were not correctly magnetized in this step.

7. Be sure no iron table parts or stray magnets are nearby to influence the compass. If necessary, raise the compass on a plastic cup or a few books to stay clear of stray magnetic fields.

If the hat rests a little lopsided, trim excess paper from the downward side.

The north hat-pin should point north (and the south hat-pin south) no matter how you turn the base of the compass. If it doesn't, remagnetize the pins as illustrated in step 6.

This hat-pin compass has considerably more turning friction than the hairline compass made in activity 8. As a result, it tends to settle down much faster. But it may sometimes need to be nudged or jiggled a little so the hat-pins will center squarely on the north-south magnetic axis.

Students should save their compasses for the next activity.

Answers

7. HOW SAME: Both are compasses – small suspended magnets that point toward magnetic north and south.

HOW DIFFERENT: The needle suspension is less sensitive than the hair, but has a full 360° of turning freedom.

MY PREFERENCE: The hat-pins compass is easier to use, since it settles down faster and you don't have to line it up. The hairline compass is more sensitive and easier to carry.

Ask your students to demonstrate agreement between the hairline and hat-pins compasses. For example, show that both agree on the direction of west.

Materials

☐ A 3x5 inch index card.
☐ Cellophane tape.
☐ Scissors.
☐ Steel pins.
☐ A magnet.

PIN MOTORS

1 Set up the hat-pins compass you made in activity 15.

2 Tape a magnet *above* the jaws of a clothes-pin. Then tape the clothespin to the table.

3 Make your compass spin around like a motor, by squeezing the clothes-pin open and closed.

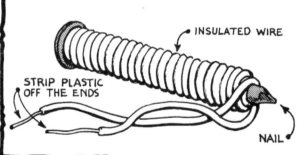

Why does the compass keep spinning?

(Continue answer on back if necessary.)

4 Use your electromagnet from activity 14, or make a new one.

INSULATED WIRE

STRIP PLASTIC OFF THE ENDS

NAIL

5 Stick the electromagnet in a small lump of clay near the compass. Stand a dry cell on one of the wires.

PUT NAIL HEAD NEAR THE PIN HEAD

6 This time make your motor spin using electricity.

off
on

Tell how your electromagnet motor works.

(Continue answer on back if necessary.)

TOPS LEARNING SYSTEMS

Objective

To build simplified models of an electric motor. To understand how they work.

Lesson Notes

3. Some practice is necessary to develop a sense of timing. By trial and error, you must determine exactly when to lever the clothespin up during each revolution, and when to let it down. When you find a way that works, note the position of the bold arrowhead printed on the hat. If you push and relax the clothespin each time this arrow head rotates past this *same position*, the hat will continue to spin.

It's not easy to explain why the hat-pins spin. Like riding a bicycle, it's easier to do than to explain the physics involved. Encourage your students to verbalize their answers first, then write down what they've said.

5. Keep the gap between the pin and the electromagnet no larger than the width of a pencil.

6. The hat-pins spin here just as they did in step 3, due to variations in the applied magnetic field. There is only one difference: In step 3 you shifted the field up and down; here you turn it on and off.

Answers

3. When I raise the magnet, it attracts one of the pins, repels the other, and begins to turn the hat pins. But before these pins stop a half turn later, I lower the magnet to let them swing on by (powered by their own inertia). When they reach the position where I originally got them moving, I raise the magnet to attract/repel them once again. By moving the clothes pin up and down, in time with each revolution of the hat pins, I keep them turning.

6. I momentarily touch the wires to the dry cell. This creates a magnetic field, attracting one of the pins and repelling the other. After the hat-pins spin one full revolution, back to their starting point, I momentarily turn the electromagnet on again, to attract and repel the pins again. With every revolution of the compass, I give it another push with my electromagnet.

Materials

☐ The hat-pins compass from activity 15.

☐ A clothespin.

☐ Tape.

☐ A magnet.

☐ The electromagnet from activity 14.

☐ A lump of clay.

☐ A size D dry cell.

DOTS AND DASHES

1 Wrap 30 cm (1 foot) of bare wire about 5 times around a paper clip.

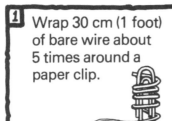

2 Tie the other end around a styrofoam cup.

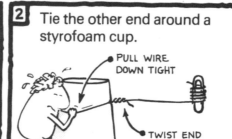

PULL WIRE DOWN TIGHT

TWIST END

3 Clamp a clothespin to the styrofoam cup, then stand it in clay.

4 Clamp the electromagnet you made before in another clothespin. Stand this clothespin in clay as well.

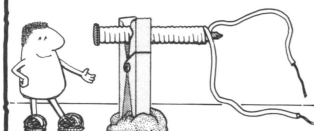

5 Slide both clothespins together so the nail head *almost* touches the copper wire wrapped around the paper clip.

VERY SMALL GAP

6 Make your telegraph click on and off.

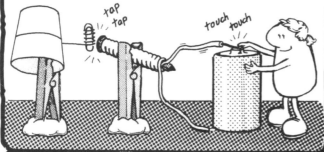

tap tap   touch touch

7 Have someone ask you questions. Answer them on your telegraph.

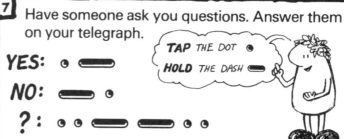

YES: ● ▬

NO: ▬ ●

?: ● ● ▬ ▬ ● ●

TAP THE DOT ●
HOLD THE DASH ▬

If the paper clip *won't move:*

➤ Make the gap smaller.

➤ Make the wire arm longer.

➤ Turn the dry cell around.

➤ Use 2 cells in series.

If the paper clip *sticks* to the nail:

➤ Be sure the nail touches only copper wire, *not* the paper clip.

➤ Wrap the copper wire 1 or 2 more turns around the paper clip.

8 Tell how your telegraph works.

(Continue answer on back.)

SAVE YOUR TELEGRAPH PARTS TO MAKE A BUZZER!

TOPS LEARNING SYSTEMS

Objective

To build a working model of a telegraph. To understand how it functions.

Lesson Notes

Activities 17, 18 and 19 all involve inventions: a telegraph, a buzzer, and a motor. Your students have the opportunity to become expert engineers in a world of paper clip technology – to twist, turn and adjust until the telegraph clicks, the buzzer vibrates, and the motor spins.

Be sure to build each invention yourself before your students try it. In this way you'll be familiar with the instructions and have a model to show your class.

Be prepared for lots of initial excitement, followed by some *"it-won't-work"* frustration, followed by *"I-finally-did-it!"* triumph.

Each activity requires some eye-hand coordination. Activity 17 is the easiest, activity 18 a little harder, and activity 19 the most difficult. If your younger, less coordinated students can't build the telegraph without considerable adult help, consider studying the other inventions in a different context, perhaps as a teacher demonstration, or as a science project for selected students.

1. The paper clip and electromagnet tend to become permanently magnetized when touched together, causing them to "stick" even after the telegraph is turned off. This problem is prevented by wrapping the clip in copper wire. The telegraph clicks as copper wire strikes the nail. The paper clip remains separated from the electromagnet by the diameter of the wire.

2. The styrofoam cup amplifies the clicking sound. To prevent this sound from becoming soft and "mushy," pull the wire as snugly as possible onto the cup.

5. The sounder must meet the electromagnet at roughly a right angle so the paper clip can freely swing toward the electromagnet and spring away.

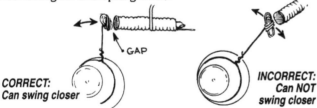

CORRECT:
Can swing closer

INCORRECT:
Can NOT
swing closer

Keep the space between paper clip and nail head as narrow as possible. Tilt the cup, turn the electromagnet, bend the wire arm, whatever it takes to make them *almost* touch.

6. The wire arm acts as a spring, pulling the sounder away from the electromagnet when you release the telegraph key. If this wire is too long, the clip may stick to the electromagnet, or double-click as it springs away. If too short, it may not be flexible enough to let the sounder move at all. Your students must adjust the wire accordingly – until the spring tension is about right.

Students who have had little experience with making things work tend to throw up their hands and run to the teacher for help at the first sign of trouble. Don't jump in too quickly. A

troubleshooting table is provided so students can solve their own problems. Those who persevere enjoy a boost of self confidence – knowing they did it all on their own!

7. To familiarize your class with Morse code, tap yes-no-maybe responses on your desk with a pencil. Sound the dots more rapidly than the dashes. Pause a little longer between letters. Ask questions of your class, and ask volunteers to tap back answers. If students have trouble distinguishing between "yes" and "no" (abbreviated A and N), spell out these words.

For those who wish to send more extensive messages, provide a copy of the supplementary activity page on International Code.

Extension

As a class project, try extending a telegraph line across your classroom. You'll need 2 long wires with a telegraph at each end.

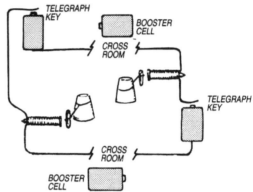

Be sure the cells at each end are connected in series (positive to negative). If your cells are not fresh, you may need to add one or two booster cells in series to overcome the increased resistance in the long wire.

Notice that the circuit forms one long, continuous loop that can be broken at either telegraph key, so in order to receive a message, you must keep your key depressed.

Answers

8. When I press the telegraph key, the electromagnet attracts the iron in the paper clip sounder, making it click against the nail. Then I release the key to stop the flow of electricity and thereby break the magnetic field. This allows the sounder to spring away from the nail.

Materials

☐ A paper clip.

☐ Bare copper wire, about 24 gauge. You may substitute aluminum, but not steel wire. Don't use wire thicker than 22 gauge.

☐ Clothespins.

☐ A small styrofoam cup.

☐ The electromagnet from activity 14.

☐ A lump of clay.

☐ A size D dry cell.

DOES IT BUZZ IT?

1 Unwind about 30 cm (1 foot) of wire from your electromagnet.

30 cm

SHORT END POINTS OUT LIKE THIS.

2 Clothespin the wire over the lip of a bottle or jar. Adjust the electromagnet so it swings free and balances level.

FREE AND LEVEL!

3 Slide your telegraph sounder (from activity 17) near the electromagnet. The nail should *almost* touch the paper clip.

ALMOST TOUCHING

4 Make a buzz. Hold long wire A to one end of a dry cell while you press the other end gently against short wire B.

A

B

PUSH LIGHTLY

5 Tell how your buzzer works.

(Continue answer on back.)

If the paper clip *won't buzz:*

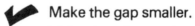 Make the gap smaller.

Make the wire arm longer.

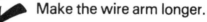 Turn the dry cell around.

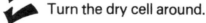 Use 2 cells in series.

If the paper clip *sticks* to the nail:

Be sure the nail touches only copper wire, not the paper clip.

Wrap the wire 1 or 2 more turns around the paper clip.

TOPS LEARNING SYSTEMS

Objective

To build a working model of a buzzer. To understand how it operates.

Lesson Notes

2. Any large, stable support – a jar, box, or ring stand – works fine. Notice that the wire loops over the lip of the support, and is clamped firmly with a clothespin on both sides.

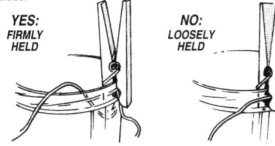

YES: FIRMLY HELD

NO: LOOSELY HELD

3. As with the telegraph, the sounder must meet the electromagnet at an approximate right angle so the paper clip can swing freely toward the electromagnet and spring away again. This back and forth motion is rapid enough to create a vibrating buzz.

4. The nail should almost touch the copper wire wrapped around the paper clip. You'll need to make fine adjustments: tilt the cup, turn the electromagnet or bend the wire arm to bring them very close together.

Students tend to hold onto the free-swinging electromagnet. If they do this, the buzzer won't buzz because the electromagnet can't vibrate. Holding the electromagnet converts the buzzer, in effect, into a telegraph.

To operate the buzzer, hold **only** the cell touching this gently against the electromagnet. This moves the electromagnet near the paper clip, causing both to vibrate. The bouncing of the electromagnet against the dry cell creates the on-again-off-again magnetic field that sustains the vibration.

Again, a troubleshooting chart is provided so that students can experience the satisfaction of solving their own "it-doesn't-work" problems.

Discussion

Here's what an electric bell might look like from the outside. Consider with your class how the insides might be wired.

BELL

HAMMER

Here is one possible design:

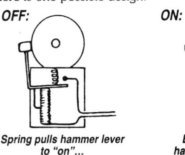

OFF: Spring pulls hammer lever to "on"...

ON: Electromagnet pulls hammer lever to "off"...

Answers

5. When the paper clip and electromagnet are apart, this completes the circuit, creating a magnetic field that attracts them together. As they click together, this breaks the circuit and turns off the field. So they spring back apart. This on-again-off-again movement back and forth creates a buzzing sound.

Materials

☐ The electromagnet from activity 14.

☐ A jar, box or ring stand to support the electromagnet.

☐ Clothespins.

☐ The sounder from activity 17.

☐ A lump of clay.

☐ A size D dry cell.

ON-OFF MOTOR

1 Cut a piece of foil as big as an index card. Fold it in half 3 times along its length.

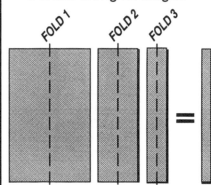

FOLD 1 FOLD 2 FOLD 3

2 Paper-punch a hole in the middle. Then cut the strip in half across this hole.

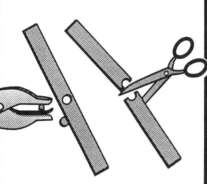

3 Slide each half under a rubber band tightly wound around each end of a dry cell. Set 2 magnets in between.

CRADLES

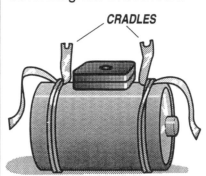

4 Coil an 8 cm piece of thin, bare wire 1⅓ times around your pencil.

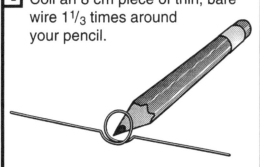

5 Straighten the "arms" so the wire spins easily, resting in the foil "cradles" above the magnet.

Press the foil leads to the battery terminals.

CONTACT!

Give the coil a spin to get it started...

6 Number these boxes in the correct order:

☐ *Electricity moves around the coil again.*

☐ *Electricity moves around the coil.*

☐ *The coil turns a half turn, repelled and attracted by the magnets.*

☐ *Press the foil leads to the terminals.*

☐ *Electricity stops flowing.*

☐ *The coil's arm bounces up as it turns.*

☐ *The coil becomes an electromagnet.*

☐ *The arm of the coil comes back down.*

☐ *The coil coasts another half turn.*

7 Describe in your own words how your motor works.

TOPS LEARNING SYSTEMS

Objective

To build a tiny motor that spins by turning itself on and off. To understand how it works.

Lesson Notes

It's hard to believe that you can build a working motor from nothing more than foil, magnets, wire, rubber bands and a battery. But you can! And it works beautifully – if you build it right. Older students can probably work independently. Younger students will need assistance, especially when forming and balancing the coil (steps 4-5).

2. Cutting across the hole creates two semicircles that cradle the wire loop in step 3.

3. The rubber bands should be tightly wound to hold the foil securely to the battery. This allows the outside leads to be touched to the battery without moving the inside supports. (Take care to keep these supports from touching to avoid shorting out the battery.)

The magnets may be magnetically attracted to the battery casing. If not, hold them in place with a loop of masking tape rolled sticky-side out.

4. Notice that $1/3$ of the coil contains a double strand while $2/3$ of the coil has only a single strand. This distributes the weight evenly above and below the wire arms, for smoother spinning in step 5.

5. Adjust the side supports as necessary so the coil turn as close as possible to the magnets without touching them. Add a clay "foot" if needed to keep the battery from rolling.

Insulated wire is normally used to wrap coils. This motor is an exception. Natural spring in the wire should keep the loop from closing upon itself and shorting the loop. (Check that this is so.) And lack of insulation keeps the coil extremely light.

Attracting/repelling magnetic fields are what drive electric motors. Most have commutators that change the direction of the magnetic field with each half turn of the coil (so it won't simply flip once and freeze). Our little motor alters the magnetic field in its coil, too, but in a much simpler way. By bouncing as it rotates, the coil turns itself on and off, alternately creating and losing its electromagnetic field. But because its bounce is random, it spins in rather amusing fits and starts, first one way and then the other, sometimes slow, sometimes fast.

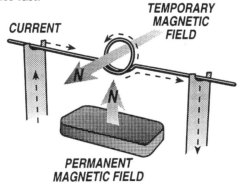

CURRENT

TEMPORARY MAGNETIC FIELD

PERMANENT MAGNETIC FIELD

Answers

6. The boxes should be numbered in this order:

9 Electricity moves around the coil again. (Then to **3**...)

2 Electricity moves around the coil.

4 The coil turns a half turn, repelled and attracted by the magnets.

1 Press the foil leads to the terminals.

6 Electricity stops flowing.

5 The coil's arm bounces up as it turns.

3 The coil becomes an electromagnet.

8 The arm of the turning coil comes back down.

7 The coil coasts another half turn.

7. When you touch the foil leads to the battery, electricity flows through the coil, turning it into an electromagnet. One side of the coil is attracted to the permanent magnets and the other side is repelled, causing it to turn. But before it arrives at a stable position and stops, the coil bounces and breaks the circuit, turning the electromagnet off. This allows the coil to swing on by before the arm settles down and reconnects the coil to be attracted and repelled again in a repeating cycle.

Materials

☐ Aluminum foil.

☐ An index card, 3 x 5 inches (optional).

☐ A paper punch and scissors.

☐ A metric ruler or meter stick.

☐ Two rubber bands.

☐ Two magnets.

☐ A size-D dry cell.

☐ An 8 cm piece of very thin, shiny, bare wire – copper or aluminum – about 30 or 32 gauge. Never substitute heavier wire; it will not work. Strip the insulation from wrapping wire, or purchase uncoated wire from an electronics store. If its surface is even slightly tarnished, polish the contacts with steel wool. To keep the motor in top running condition, remove all traces of oxidation from time to time, and freshly trim the aluminum supports.

☐ A lump of clay (optional). See note 5 above.

RICE ROUNDUP

1 Place an index card lengthwise at the center of half a manila folder.

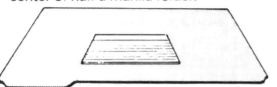

2 Draw circles that touch each end by tracing around a dry cell. Label them "A" and "B".

THEN REMOVE THE INDEX CARD

A

3 Tape 2 clothes hangers near the edges of the folder.

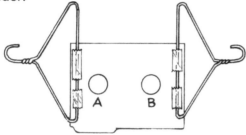

A   B

4 Rest books on the hangers so the folder can hang out over the edge of your desk like a platform.

Tape the edge of the folder to the edge of your table.

5 Put 12 rice grains in circle A and an **open** staple in circle B . . .

START...

A   B

Then have a rice round-up! Herd all the rice grains from A to B, *using a magnet.*

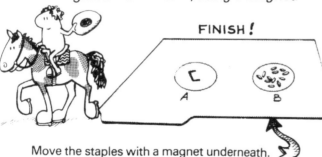

FINISH!

A   B

Move the staples with a magnet underneath.

6 Here's another challenge. Start with 4 rice grains and 4 open staples in the middle. Using only your magnet, arrange them like this.

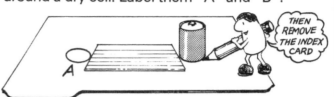

A   B

A   B

START...   FINISH!

7 Are your staples magnets, or simply magnetic? Explain your reasoning.

(Continue answer on back.)

TOPS LEARNING SYSTEMS

Objective

To enjoy a game of skill using magnets.

Lesson Notes

1-2. The index card and dry cell standardize the size of the circles and distance between them. Later in step 5 you may wish to organize a class competition to see who can round up rice grains fastest. So it's important that the corral areas have a uniform size and position.

3. The clothes hangers shouldn't get too near either circle. Otherwise the iron in the hangers may grab the staple as you move it about with your magnet beneath.

4. The manila platform should rest level and flush with the table. If it sags or ripples, try adjusting the tension on the clothes hangers by shifting the books that weigh them down.

5. The orientation of the magnet you hold below the board is important. If the poles point out, then the staple will be easy to control. If the poles point up and down, staple control is difficult. You might show younger students how to hold their magnets, while letting your older students puzzle this out for themselves.

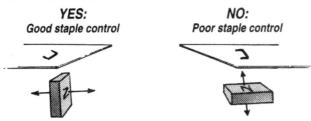

YES:	NO:
Good staple control	Poor staple control

Just for fun, hold a class contest. See who can be first to get from start to finish, moving all 12 rice grains with only the staple and magnet. Your class will enjoy the competition.

6. The staples tend to stick together due to mutual magnetic attraction. Part of the challenge is trying to separate them and keep them apart. This is easier to do with two magnets, although it is possible with only one. It's no fair touching the staple or rice grains with your fingers.

Extension

Challenge your students to make their staples dance! To do this they should hold the magnet so its poles point vertically. By bringing the magnet close enough to the cardboard (not too close), like poles will repel so that one end of the staple stands vertically. Gentle movements of the magnet will cause the staple to sway to and fro.

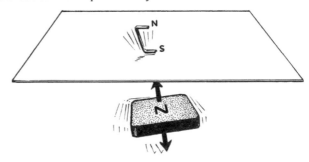

Answers

The staples are magnets. If I bring two staples together, they will attract and repel like magnets. When I turn the large magnet below, the staples above also turn, maintaining a preferred orientation with unlike poles attracting and like poles repelling.

Materials

☐ A 3x5 index card.

☐ A manila folder cut in half. Or substitute a similar piece of thin cardboard.

☐ A size D dry cell.

☐ Clothes hangers. Plastic rulers or similar long flat supports will also serve.

☐ Tape.

☐ Books or other weights to ballast the clothes hangers.

☐ Uncooked rice grains.

☐ Unbent staples.

☐ A magnet.

SUPPLEMENTARY PAGES

ATOMS

for MAGNETIC MODELS

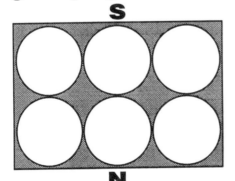

Draw circles and arrows:

a. nonmagnetic aluminum

b. denmagnetized iron (arrows cancel)

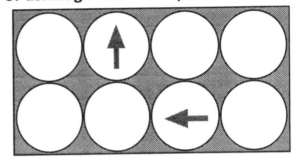

c. nonmagnetic copper

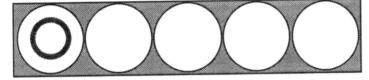

d. strongly magnetized iron

N 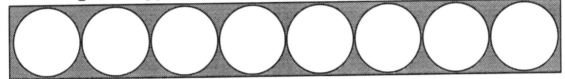 **S**

e. demagnetized iron

f. weakly magnetized iron

N 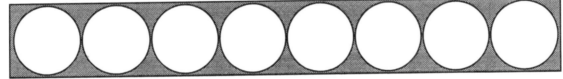 S

g. strong magnet:

h. medium magnet:

i. weak magnet:

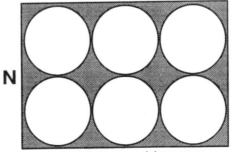

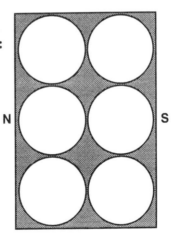

INTERNATIONAL CODE
(A Standardized Form of Morse Code)

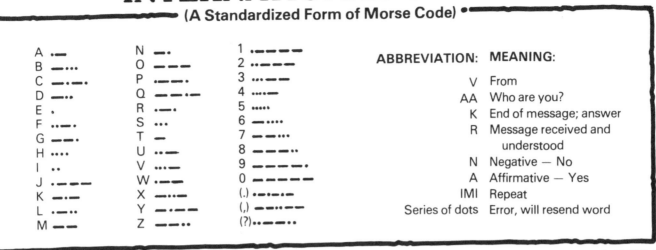

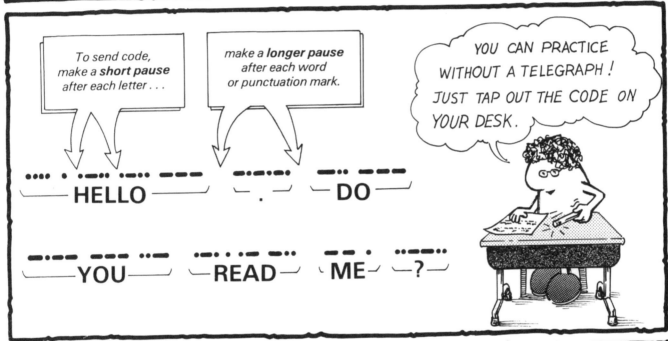

Feedback!

Dear Educator,

If you enjoyed teaching TOPS, please say so. Your praise will motivate us to work harder. If you found an error or can suggest ways to improve this program, we need to hear about that too. Your criticism will help us improve our next new edition. Do you need information about our other publications? We'll send you our latest catalog free of charge.

For whatever reason, we'd love to hear from you, and will carefully consider your input. We include this self-mailer for your convenience.

Sincerely,

Ron & Peg

Ron and Peg Marson
author and illustrator

module title _____ date _____

name _____

address _____

city _____ state _____ zip_____

1st FOLD   TAPE CLOSED

2nd FOLD

_____

_____

_____

TOPS Learning Systems
10970 South Mulino Road
Canby OR 97013